MANUFACTURING ADVANTAGE

STUDIES IN EARLY AMERICAN ECONOMY AND SOCIETY
FROM THE LIBRARY COMPANY OF PHILADELPHIA
Cathy Matson, SERIES EDITOR

MANUFACTURING ADVANTAGE

War, the State, and the Origins of American Industry, 1776–1848

Lindsay Schakenbach Regele

JOHNS HOPKINS UNIVERSITY PRESS
Baltimore

Printed in the United States of America on acid-free paper
9 8 7 6 5 4 3 2 1

Johns Hopkins University Press
2715 North Charles Street
Baltimore, Maryland 21218-4363
www.press.jhu.edu

Library of Congress Cataloging-in-Publication Data

Names: Schakenbach Regele, Lindsay, 1984– author.
Title: Manufacturing advantage : war, the state, and the origins of American industry, 1776–1848 / Lindsay Schakenbach Regele.
Description: Baltimore : Johns Hopkins University Press, 2019. | Series: Studies in early American economy and society from the Library Company of Philadelphia | Includes bibliographical references and index.
Identifiers: LCCN 2018020730 | ISBN 9781421425252 (hardcover : alk. paper) | ISBN 1421425254 (hardcover : alk. paper) | ISBN 9781421425276 (electronic) | ISBN 1421425270 (electronic)
Subjects: LCSH: Industrial policy—United States—History. | Defense industries—United States—History. | Textile industry—United States—History. | United States—Politics and government—1783–1865.
Classification: LCC HD3616.U46 R414 2019 | DDC 330.973/05—dc23
LC record available at https://lccn.loc.gov/2018020730

A catalog record for this book is available from the British Library.

Special discounts are available for bulk purchases of this book. For more information, please contact Special Sales at 410-516-6936 or specialsales@press.jhu.edu.

Johns Hopkins University Press uses environmentally friendly book materials, including recycled text paper that is composed of at least 30 percent post-consumer waste, whenever possible.

CONTENTS

SERIES EDITOR'S FOREWORD

In this addition to the series Studies in Early American Economy and Society, a collaborative effort between Johns Hopkins University Press and the Library Company of Philadelphia's Program in Early American Economy and Society (PEAES), Lindsay Schakenbach Regele explores two unlikely partners in the early North American republic's rise to prominence. Together, she argues, the arms and textiles industries became the foci of policy makers' efforts to create national security in the fragile post-Revolutionary years.

Manufacturing Advantage: War, the State, and the Origins of American Industry, 1776–1848 weaves together the ambitions and institutional strategies of statesmen and manufacturers behind the production of guns and cloth, finding them at the crux of the infant American state's efforts to protect its citizens, institutions, and resources from perceived internal and external enemies, one based on military preparedness and the other based on the search for national economic self-sufficiency. Unlike the generations of colonial dependence on the power of the British Empire to protect its subjects and guide expansion, the new federal policy makers and intellectual leaders were clear that their goals depended on safeguarding political independence before surging forward into the future. Until the Revolutionary War had been won, North Americans did not have to decide what to produce or how their commerce would influence foreign affairs. But within the infant republic, the future of its newly created citizens and immigrants was at risk from persistent tax rebellions in the new states, warfare on western frontiers, and ceaseless hostilities in the Caribbean and in Europe. To overcome these and other problems, argued key officials in state and federal governments, a steady supply of reliable firearms and semi-durable clothing would be vital steps. Rebuilding a permanent dependence on the British Empire for these goods was out of the question in the new ideological framework of independence.

Schakenbach Regele tells us a compelling story about how manufacturers and government officials collaborated to transform the newly independent country from agricultural supplier to industrial producer, from colonial dependent to substantial power in Atlantic commerce. To the extent that interested parties were successful in these endeavors, it demonstrated that the United States was born from a partnership between government intervention and private initiative. Private arms production depended on direct federal funding, while entrepreneurs in the textile industry mobilized private capital with the aid of the government's friendly banking, commercial, technology, and labor legislation from the relatively early years after the Revolution. In both cases, personal ingenuity and risk-taking surely factored into success, but these private businesses also depended on federal subsidies and protectionism, trade regulations, and diplomacy to create superior economic environments. Ideologically, free-market capitalism was a non-starter; Schakenbach Regele insists that "market competition and capital investment for the sake of profit were secondary to economic and military independence and protection." In short, what we recognize as a regular feature of the North American economy by the late nineteenth century—the partnership of private and public capital in the interest of national and public security—had its roots in the early republic.

Cathy Matson
Richards Professor of American History, University of Delaware,
and Director, Program in Early American Economy and Society,
Library Company of Philadelphia

ACKNOWLEDGMENTS

I am not sure how to thank properly the many individuals and institutions who have generously given time and resources to help me write this book.

This project began during my second semester at Brown University in Michael Vorenberg's legal history seminar. I am forever grateful to him for his early guidance and generous feedback. Brown University was a wonderful place to be a graduate student, and the early phases of this project benefited from the encouragement of many faculty and fellow graduate students, especially Laura Perille, Ania Borejsza-Wysocka, John Rosenberg, Ben Holtzman, Liz Searcy, Sara Mattiesen, Anne Gary Fischer, Eunsun Han, Zack Dorner, Patrick Chung, Rachel Knecht, Henk Isom, and Alicia Maggard. Hal Cook was a generous reader of my work, and the conversations I had with him helped me see the early United States through the lens of early modern commercial networks. I owe my biggest intellectual debt to my adviser, Seth Rockman. He changed how I think about the past, shaped the development of my project, and cleaned up my messy writing, all while making me think that it was my own doing. I cannot thank him enough for balancing critical pen marks on chapter drafts with kind words of encouragement. Every graduate student should be so lucky to have a Seth Rockman in her or his corner. He continues to be a wise, witty, and generous mentor.

A number of other institutions provided the community and resources necessary to complete this work. In addition to Brown University and Miami University, Harvard University, the Colonial Dames Society, the Rovensky Fellowship, the Massachusetts Historical Society, American Antiquarian Society, the Library Company of Philadelphia, and the Huntington Library provided funding for archival research. The extraordinarily knowledgeable and helpful staffs at the Waltham, College Park, and Washington, DC, branches of the National Archives; the American Antiquarian Society; Harvard Business School's Baker Library; the Historical Society of Pennsylvania; the Library Company of Philadelphia; the Library of Congress; the Connecticut, Rhode Island, Massachusetts, and Middlesex County Historical Societies; Yale University; and the Huntington Library answered questions and offered research leads and advice with patience and enthusiasm. I would like to extend additional thanks to Conrad Wright, Kate Viens, Paul Erickson, Caroline Sloat, and Connie King, and the generous McNeil community in Philadelphia.

The opportunities to present portions of this book to helpful audiences at the Massachusetts Historical Society, American Antiquarian Society, the Yale Early American Historians workshop, the Library Company of Philadelphia, the Ohio Seminar in Early American History, as well as at conferences at Harvard and Cornell, and at the annual meetings of the Society for Historians of the Early American Republic, the Business History Conference, the Society for Historians of Technology, and the American Historical Association improved the final outcome. I had the good fortune to meet Mark Wilson during my first presentation at the Business History Conference, where he kindly agreed to serve as an outside reader for my dissertation. My work has benefited enormously from his scholarship and expertise, and he continues to serve as a professional role model.

Many individuals spent precious time offering advice and reading portions (or all!) of this book. I am indebted to Emilie Connolly, Wietse de Boer, Merritt Roe Smith, Joanna Cohen, Zorina Kahn, John Larson, Naomi Lamoreaux, Richard John, Gautham Rao, Lawrence Peskin, Brendan Gillis, Michael Blaakman, Alyssa Reichardt, Joanne Freeman, Michael Hattem, Ryan Hall, Andy Shankman, Edward Pompeian, Wendy Woloson, William Childs, Nora Slonimsky, Lukas Rieppel, Sharon Murphy, Margaret Newell, Mary Cayton, John Brooke, Cameron Shriver, John Belohlavek, Brian DeLay, and the late Pauline Maier. I am particularly appreciative of the helpful comments I received from Sara Damiano, Andrew Fagal, and Andrew Offenburger. I was lucky to overlap with Sara for several months in Philadelphia; I am in awe of her kindness and perceptive feedback. Andrew Fagal is the gun expert (in a not-scary way) I wish I could be, and his attention to detail is unparalleled. I am grateful to Andrew Offenburger for serving as a sounding board throughout the final stages of the writing process. We started working at Miami University at the same time and are lucky to be part of a history department that is unbelievably supportive and collegial. They have all offered encouragement on this project. The entire Miami community makes Oxford, Ohio, a great place to live and work. I pinch myself every day, but I have not woken up yet. I arrived just as Drew Cayton left, and I feel his presence every day in the halls of Upham. His legacy of generous teaching, scholarship, and friendship is a model.

For above-and-beyond hospitality and support throughout the research process I am grateful to Debra and Bill Breski, John and Linda Schakenbach, Ania Borejsza-Wysocka and Chris Bender, Kiki Bolender and Chuck Capaldi, Kendra Leith and Matt Richardson, Mary McCudden and Mike Monteleone, Carly Holbrook, Molly Palmer and Drew Colati, Dan Polifka and Katie Kalafut, and Michele and Gene Navakas.

I am enormously indebted to Cathy Matson and the editorial staff at Johns Hopkins University Press, especially Lauren Straley, Laura Davulis, Kyle Kretzer, and Juliana McCarthy, as well as copyeditor David Goehring.

I probably would not have pursued a PhD had it not been for my undergraduate adviser Leo Garofalo, who encouraged me to become a history major, and Ben Carp, who provided invaluable mentorship and feedback on my master's thesis at Tufts University. I am grateful for the support of Reed Ueda and the rest of the faculty and staff there.

And then, of course, there is my family. I could never repay my parents for their unwavering love and support and for the happiness they gave me growing up. My mother's endless patience and pride sustained me during the most difficult times. My dad was a sounding board throughout the entire process. He suggested things that I initially resisted, only to realize later that, as always, he was right. Evan is the kindest brother a sister could ask for. And Lisa, Dennis, Tim, and Bryan: thank you for everything over the past two decades.

Finally, Matt: you've made everything since eighth grade homeroom infinitely better.

MANUFACTURING ADVANTAGE

Introduction

THE FIFTY-SIX MEN who signed the Declaration of Independence agreed that the new United States had the power to "levy war, conclude peace, contract alliances [and] establish commerce."[1] This power to wage war and enter meaningful diplomatic and economic relationships depended on a level of industrialization hitherto unknown on the North American continent. In the years following independence, statesmen and manufacturers positioned firearms and textiles—two products that had been around for many centuries—to become hallmarks of American industry.

The stories of the arms and textile industries usually are not told together. The latter calls to mind Yankee capital, immigration, and female mill workers; the former, violence and government contracts. The development of the American textile industry is celebrated in grade school textbooks; the arms industry, by promoters of gun rights. In a single narrative, however, guns and cloth emerge as products of national security. National security refers to a state's ability to protect its citizens, institutions, and resources from internal and external enemies. This security hinged on military preparedness and economic stability, two factors that would allow the nation to quickly mobilize for warfare, regardless of global events.[2]

In the late eighteenth century, the only precedents for this security were colonial-era procedures, which posed problems for a new federal government whose survival rested on its ability to safeguard political independence. For almost two hundred years, the British Empire had coordinated military supply and orchestrated solutions to economic shocks. Ideological and economic complaints aside, North American colonists did not have to decide what to produce or how their commercial offerings would influence diplomacy. The non-importation movements of the 1760s and 1770s changed this to some extent, but it was not until the Treaty of Paris in 1783 that the practicalities of state-building overshadowed ideological motivations for independence. The new nation faced tax rebellions, Indian[3] warfare, and hostilities with France and England with an arsenal of British hand-me-downs from the Seven Years' War and beat-up French firearms that had been purchased on loan.[4] Without a readier supply of guns, the new nation could not defend itself; without textiles, it was at the economic mercy of the British Empire. Without either, the United States risked a recurrence of its supply dependence during the Revolution, when it had imported more than half its munitions and textiles.[5] Ideological concerns about dependence had real ramifications: in an era when armies literally froze in the field, military preparedness required not just firearms and food but blankets and jackets. Americans knew how to grow and slaughter their own sustenance, but substantial manufacturing posed more of a challenge. It was a challenge worth overcoming, however, because even in times of peace, firearms offered the promise of physical safety, while textiles provided a subtler form of security—as visible and tactile manifestations of independence.

What follows is a story of how manufacturers and government officials maneuvered the United States from agricultural supplier to industrial producer, from colonial dependent to military power. In a single narrative, the arms and textile industries allow us to see the messy distinction between government intervention and private initiative in a new nation struggling to create a political economy that balanced military competence with commercial needs. The arms industry depended on federal funding; the textile industry did not. Both industries, however, developed in a system of political economy best described as national security capitalism: a mixed enterprise system in which government agents and private producers brokered solutions to the problems of international economic disparities and war.[6] The government policies that bolstered this system tended to favor more explicitly the arms industry, because of its obvious ramifications for physical safety, but they nonetheless engendered the development of textile manufacturing as well.[7] As the manufacturers and military participants at an 1812 celebration toasted, "The best mode of warfare for our country: the artillery of carding and spinning machinery and the musketry of shuttle and sledge."[8]

National security capitalism required private capital, ingenuity, and risk-taking to sustain economic development and preparedness for military conflict. Private business likewise depended on federal resources, trade regulation, and diplomacy to generate profit. What made this different from free-market capitalism was that market competition and capital investment for the sake of profit were secondary to economic and military independence and protection.[9] Under national security capitalism, federal subsidies and protectionist tariffs enabled the United States to withstand military and economic conflicts, while "favored-nation" commercial agreements appeased domestic merchants and foreign trade partners.

Federalism limited to some extent the role of the national government in economic development: individual states were responsible for business incorporations, bank charters, and most infrastructure projects.[10] A geopolitical lens, however, reveals federal officials as central players in the development of domestic industry. The executive branch, in particular, made national security capitalism work. Although the economic prosperity component of national security can be attributed in part to state-level incorporations, banking, and infrastructure, as well as to congressional legislation on tariffs, slavery, and interstate commerce, it was the federal executive that directed military resources and conducted diplomacy in the service of industrial goods. In turn, industrial development bolstered federal power by generating revenue and martial prowess.

This symbiosis made the early national United States an example of a military-industrial complex, although the term refers to the post–Second World War United States, not the post–Revolutionary War one. Adopted by President Dwight D. Eisenhower in the 1950s as a warning against the influence of a large military establishment on society, it names the self-perpetuating connections among the military, its contractors, and congressional appropriations.[11] Defense spending in the middle of the twentieth century constituted roughly half of federal expenditures, but we can trace the genesis of the relationship between security and industry to the nation's founding.[12] Throughout the early nineteenth century, the military absorbed one-third of peacetime outlays.[13] The modern American economy, predicated on national security and the confluence of private and public capital, has its roots in the early republic.

These roots were the beginning of industrial capitalism in the United States.[14] Neither "industrial" nor "capitalism" could yet be used to describe an economy that was still largely agrarian and "moral"—indeed, "capitalism" would not enter the lexicon until the latter half of the nineteenth century—yet in the decades following the Revolution, manufacturers, capitalists, and state-builders planted the seeds for growth in the scope and scale of factory production.[15] The United States has never had an official industrial policy, as compared to say, China; state-sponsored industrial capitalism, however, was squarely within the realm of an early republican political economy predicated on national security.[16] The federal government had a compelling interest in developing national industrial resources, and its policies lurk behind the rise and fall of American factories. As producers and public officials made decisions about real-time problems, they developed ad hoc policies that would eventually propel the United States to its status as an industrial superpower. Industrial development occurred in fits and starts, but throughout the nineteenth century manufacturing processes became increasingly mechanized as capital and labor concentrated in factories and new technologies changed human relationships to work, food, consumer goods, their environments, and other people. The private ownership of factories and machines was not a radical break with a state-directed mercantile past or a harbinger of an unfettered free-market future.[17]

Instead, these gradual transitions marked the United States' taking its place among the Old World nations it considered its peers.[18] And just like the imperial governments of Europe, the new US government made decisions about production, security, and wealth—decisions that were bolstered and constrained by private producers—to enable it to fight and trade on the most favorable terms possible.[19] Historians have debated the extent to which Americans subscribed to economic liberalism or mercantilism.[20] They could simultaneously follow both models because security superseded theories and ideals. America's industrial capabilities developed in response to the geopolitical demands of the Age of Revolutions, the Napoleonic Wars, and countless named and unnamed Indian wars, rather than to abstract economic philosophies.

At the time of the Revolution, industrialization and hegemony were a long way off. Manufacturing meant making goods in households and small workshops. Few firms were incorporated, small proprietors prevailed, and "capitalist" did not yet describe the economy.[21] A major issue for the nation's founders was the scope and scale of government intervention in that economy.[22] They declared independence just as Adam Smith began to criticize the "mercantile system" of Europe, by which he meant the economic policies of central governments that competitively regulated trade and production in the service of national wealth. Even as the relationship between government and the economy was everywhere in flux, American statesmen took for granted that the government would have some role in the economy.[23] Their ideals, in fact, required an interventionist state to provide security and access to markets. James Madison, for example, championed an expansive republic whose government bolstered national welfare with territorial acquisition. Jefferson believed nations should derive their wealth from the land, and he flirted with the idea of free trade, but even he realized that an agrarian nation required commercial outlets for its produce and home manufacturing to sustain frontier life—both of which functioned best with some government intervention.[24] And Alexander Hamilton advocated the most activist role for a central government that directed public finance and protected domestic manufacturing.

Of the three branches of government laid out in the Constitution, it was the executive that would have the most significant effect on economic development. A little over a century after the creation of the US government, Max Weber described as inevitable the bureaucratic inefficiencies that result from governmental separation of powers.[25] There was, however, sufficient "energy in the executive," as Alexander Hamilton had recommended for effective governance, to permit the federal promotion of US industry.[26] The executive branch had the power to shape industry in a way that Congress did not. Congress made laws that the judiciary upheld or challenged, but protective measures for manufacturing interests often got hamstrung by legislative hurdles.[27] The executive branch, on the other hand, doled out public monies and supplies and conducted business with foreign governments. Congress declared war, but the executive *made* war.[28] Article Two, Section Two, of the Constitution gave the president the power to appoint cabinet officers who would oversee war, the economy, and foreign affairs. Accountable to Congress, but unencumbered by voting constituents, these officers made it possible for the federal government to create a federal arms supply and support a textile industry that could compete with British imports at home and, eventually, abroad.

The executive was most effective in the realm of international relations, which were central to the early republic, rather than being, as many diplomatic historians used to contend, "nascent" or "isolationist."[29] Early American foreign policy, while not geared toward large-scale military engagements as it was in the twentieth century, shaped the international relationships that made US economic expansion possible. Federal officials and manufacturers used the state's nation-building capacities to generate economic opportunities at home and abroad.[30] The State Department negotiated treaties that secured favorable trade policies and land acquisition, and it employed growing American power to open new markets for domestic manufactures overseas. In particular, it looked to Latin America, a region on which the United States projected both its anxieties and its hopeful opportunism about nation-building.[31]

The state's ability to manage markets at home and abroad depended on the ability to mass-produce weapons that could be traded in peacetime and fired in war. The early United States, in fact, looked a lot like a reincarnation of the resource-gathering, surplus-avoiding, and military-wielding behaviors of European mercantile states of the seventeenth and eighteenth centuries, even as many Americans feared the large standing armies of Europe and believed state militias offered the best protection for republican citizens.[32] Their fears of militarism, however, lost out to violent conflicts with Indian nations over land and threats of warfare with Europe, which necessitated greater armed forces than local militias could provide.[33] As historian Max Edling has shown, the early national United States was in no way exempt from the realities of war making. The ratification of the Constitution, in fact, called into being a fiscal-military state à la the European model.[34] The military was essential to the new nation-state. It made warfare and industrial development possible.

Historian C. Vann Woodward referred to the one hundred thirty years between the end of the War of 1812 and the Second World War as an era of "free security" in American history, in which Americans enjoyed safety with minimal military spending—generally less than one percent of the gross national product.[35] While political scientists and foreign policy scholars have appropriated the term to describe America's role in international affairs, it overstates the existence of peace—especially as the United States was always engaged in warfare against Indians—and obscures both the real costs of maintaining a military and the effects of this military on the economy.[36] If the years following the second war with Britain seemed "free," it was only because the previous twenty-five years had been spent building an effective military organization and developing production capabilities.[37]

Following war with Britain, this "free security" hinged on the strategic consolidation of land for white settlers and slave-grown commodities. Land, though, was not free. Whether we consider large portions of the continent a "frontier" or "borderlands," the reality was that land was contested and warfare was perpetual, or at least perpetually possible.[38] Beginning in the 1810s, the United States was involved in an almost never-ending war in the southeast, along with fighting throughout the Great Lakes region, as well as official Indian removal in the 1830s. As the Jeffersonian ideal of an agrarian republic collided with the broader national phenomenon that would become known as Manifest Destiny, the State and War Departments provided complementary services for American manufacturers and national safety.[39] From the 1810s through the 1840s, the United States added Florida, Texas, California, and other territories to its Louisiana holdings. The State Department hammered out deals that clarified territorial boundaries and improved prospects for American manufactured exports in countries from Peru to China. The War Department organized frontier defense and carried out war against Mexico.[40] The sum total of their efforts provided outlets for machine-produced muskets, rifles, and coarse woolens for military use, and sheetings, shirtings, jeans, and ginghams for civilians. (Textile terms are defined in appendix A.)

While industrial development was intricately linked with military security and economic independence, the influence of federal policy played out differently in different locations, just as the transition to capitalism was not a monolithic process.[41] We must understand industrialization as more than labor management styles or levels of capitalization. There was more at play than whether a firm relied on factory laborers or household workers, or whether it was funded by local merchants or capitalists who lived a train's ride away, or whether it was capitalized at $20,000 or $200,000. For when we step back, we see that different industrial realities arose from a firm's position within a constellation of national and international factors, such as the extent to which military conflict generated demand, where and to whom contracts were given, and the channels through which one had access to diplomatic negotiations abroad.

A factory's location within the international economy started with its geographic position.[42] By the middle of the nineteenth century, New England began to develop its historical reputation as the birthplace of industrialization in the United States. Large, single-industry towns that produced such goods as fabrics, firearms, carpets, clocks, shoes, and precision machines began to dominate the landscape. Eastern New England became known for its textile production, and the Connecticut River Valley as a hub of premier gun manufacturing. By 1831, the amount of capital invested in cotton establishments in New England surpassed that of anywhere else in the country. Massachusetts capitalists invested over $12 million in cotton factories and Rhode Island over $6 million. Outside of New England, the state with the largest cotton industry, Pennsylvania, had less than $4 million tied up in cotton establishments.[43] New England was also where, beginning in the 1810s, the federal government acquired 85 percent of the small arms with which it supplied state militias.[44] These statistics are why the subject of New England and the American Industrial Revolution has received so much attention since the early twentieth century.[45]

The banality of New England industrialization almost makes it unworthy of further study.[46] Significant industrial development and government contracting occurred elsewhere, most notably in the Mid-Atlantic, and often predated New England's. Philadelphia's industry, for example, was characterized by a small-scale, flexible production system that manufactured high-quality specialized goods, as well as coarse cloths and weapons for the military.[47] The region was also home to family-based manufacturing operations that made small investments in separate spinning and weaving operations and generally opposed incorporation.[48] Scholars have spilled a lot of ink to refute the idea of New England as emblematic of American industrialization, arguing either that its factories were atypical in their high levels of capitalization, or that they lagged behind their Mid-Atlantic counterparts in industrial innovation.[49]

All of this was true, but New England industrialization was more than the Samuel Slater or Lowell Mill Girl paradigms. As a region, it ran the gamut from large factory town to tiny mill village, from factory owners who claimed they wanted minimal protection to those who begged for government assistance. It was where sustained, large-scale integrated textile manufacturing eventually developed first, but it also was where aspiring manufacturers shut their mill doors every time the nation experienced an economic shock. Farther west, the Connecticut Valley formed the core of federal arms production, even though the best firearms of the colonial era were made in the Mid-Atlantic. War Department officials did more to standardize and expand private arms manufacturing in western New England than they did in the region surrounding the other federal armory at Harpers Ferry, Virginia.[50] Their influence explains why, by the mid-nineteenth century, the biggest patent firearms companies were in New Haven and Hartford, Connecticut, rather than in Lancaster, Pennsylvania, which in the eighteenth century was a hub of high-quality rifle production.

In many ways the major movers in these industrial regions were not investors or the manufacturers themselves, but a cast of federal officials: secretaries of state, treasury, and war, ordnance officers, and consular agents, who issued contracts and patents, conducted manufacturing reports, and employed diplomacy to benefit American industrial exports. They worked on behalf of both public policy and private sector interests. New England's arms and textile industries reveal the mobilization of private capital and government resources in the service of economic expansion and hemispheric dominance. For these reasons, it is the ideal place to study the relationship between national security capitalism and the transition from handcraft manufactures to factory production. It can also help us understand the relationship among the political center, the geography of producers, and the territorial periphery of the United States.

The influence of the government varied throughout New England, as its three major rivers—the Blackstone, the Merrimack, and the Connecticut—occupied different places in the national and international systems of production. The Blackstone Valley and Merrimack Valley were the centers of American textile production while the Connecticut Valley was the capital of firearms. From the Revolution up through the Mexican-American War, the balance between public policy and private initiative fluctuated in relation to manufacturing. The federal government channeled most of its energies into the arms factories throughout the hinterlands of the Connecticut River Valley, from Pittsfield and Springfield, Massachusetts, down to Middletown, Connecticut. It was less interventionist with the textile industry; rather than invest in new enterprises, it offered indirect assistance that privileged more highly capitalized ventures. For both industries, however, the level of support ebbed and flowed depending on geopolitical circumstances.

The Revolutionary War brought to the fore the difficulties in providing clothing and weapons for soldiers. Although the war highlighted the need for domestic military-industrial capabilities, efforts were decentralized and reactive, and following the war a sort of historical amnesia set in among policy makers and consumers as Americans welcomed the resumption of imports, primarily from the British Empire. This was especially true for textiles, whose import from England increased following independence.[51] If this trade relationship continued, the United States would remain in the shadow of European empires, struggling to establish economic and military security in an insecure world. In the face of border conflict, slave uprisings, and tax rebellions, this inadequacy proved a grave threat. Many American policy makers, however, did not want to turn their nation of small farmers into a nation of industrial capitalists or wage earners, certainly not one with a large standing army. What would the solution be? Despite misgivings about the social ramifications of large-scale industry and militarization, leaders like George Washington recognized that the surest way to preserve peace was to be prepared—with arms—for war. Guns were of course more of a military imperative than textiles, but textiles mattered too, especially as visible manifestations of independence. Policy makers began to take some initiative to improve domestic manufacturing. Federal policies, analogous to Keynesian stimulus spending, generated demand for domestic arms and textiles in a way that private consumption did not. In the 1930s, British economist John Maynard Keynes postulated that consumer demand sometimes fails to reach equilibrium with the supply capabilities of producers. The solution, he argued, is government expenditure.[52] The early national federal state provided this solution by growing the market for firearms and by helping textile manufacturers break into the British-dominated market.[53]

After the Revolution, the federal government committed to creating a program for national arms production. The US government founded two federal armories at Springfield, Massachusetts, and Harpers Ferry, Virginia, and developed a plan for arming the militia. It also began experimenting with contracts for muskets and rifles, of which the latter had been used mostly on the frontier for hunting and the former imported from Europe for colonial warfare. The federal government developed the Connecticut River Valley as a technology district akin to today's Silicon Valley, where it oversaw production at the armories and established the beginnings of decades-long relationships with several arms makers near the Springfield Armory.

The textile industry would never get the sort of centralized support that the arms industry received. In general, policy uncertainty prevailed for the nation's first quarter century—the first significant protective tariffs for cotton and woolen goods were not enacted until 1816—but there were efforts to stimulate domestic production. Beginning in 1806, the federal government limited imports of woolen cloth and ready-made clothing from Great Britain and began soliciting samples from domestic manufacturers for potential military use. During the War of 1812, it raised prices to entice factory owners to switch from civilian to military production. It also contracted out for domestically produced army cloth and arranged for the making of army uniforms at the federal supply depot in Philadelphia. In 1815, the United States emerged from a war with a major world power in which its soldiers had fought with American guns in their hands and at least some domestic textiles on their backs. The supply nightmares of the Revolution were over.

With manufacturing no longer a question, American policy makers asked how best they could channel federal resources into public and private industries to ensure national prosperity and dominance over the continent. In the context of the acquisition and defense of Florida, the government stimulated industrial improvements in both the arms and textile industries. The claims payments that a group of Boston merchant-industrialists received as a result of the 1819 Transcontinental Treaty with Spain, whereby the United States "acquired" Florida, served as financial stimulus for their textile manufacturing ventures. Allowing recipients to buy out smaller enterprises, these claims payments helped make possible industrial consolidation in the Merrimack Valley. As the United States waged war against Seminole Indians in Florida and required greater frontier protection, it increased federal funding and bureaucratic oversight to grow and innovate the arms industry. War Department officials imposed production standards on their contractors and worked around congressional spending limits to ramp up American weapon output. All of these federal financial strategies helped grow and modernize US industry from the late 1810s to the 1830s.

Federal officials further stimulated industrial development by helping manufacturers gain entrance into new export markets, particularly in Latin America. During the region's Independence Wars in the 1810s and 1820s, the United States tested its diplomatic and commercial might by sending federal officials to negotiate trade policies favorable to American manufactured goods. Prior to the establishment of formal diplomacy, US consuls cultivated amicable relationships with patriot leaders and facilitated the sale of weapons and other goods; following the war, agents for the United States lobbied aggressively for, and achieved, trade policies that privileged US manufactures and facilitated America's transition from industrial backwater to rising global producer.

Once manufacturing seemed more secure, the US government deployed military force in Mexico and the Pacific Northwest to consolidate federal control over the continent. The territorial possibilities of Manifest Destiny meant little without physical conquest. These efforts tested the federal government's fortitude for war making generally and industry-building in particular. The War Department had spent half a century nurturing the arms industry and, to a lesser extent, the textile industry, and that support served American expansion well. During the 1840s, War Department officials determined that the output from the federal armories largely met national needs; as a result, arms contractors no longer received large cash advances and long-term contract renewals. Some regular contractors suffered, but by that time, decades of public support had led to continuous growth in the industry, giving rise to a new class of patent arms companies. Ample demand from foreign countries and the frontier promised to sustain them. The military still contracted intermittently with textile manufacturers for specific goods like blankets and coarse woolens, but those manufacturers were on sound enough footing to withstand geopolitical shifts.

Despite variations in levels of government support, by the early 1840s US textiles beat out British ones in some foreign markets and the army outfitted its troops with uniforms and blankets made entirely of domestic manufacture—proof that textiles were not only visible signs of material independence from Britain, but physical forms of military security. The Springfield Armory was on its way to becoming the largest in the world, and federal firearms were produced with cutting-edge machine technology. Activist commercial diplomacy, endless unofficial war against Indian tribes, and an official war against Mexico provided continued sales opportunities for manufacturers.

Once neither industry depended so heavily on subsidized production, the government changed its relationship with the private sector. It swung toward "free trade" in the middle of the 1840s by slashing tariff rates. It also abandoned many of its private arms contractors in favor of more flexible manufacturing policies that included short-term contracts with the new patent firearms makers, like Colt and Smith and Wesson, who would eventually become household names. The US government did not, however, give up control of federal arms production. In 1842, it shifted management of the federal armories from civilian to military superintendence and increased the military's reliance on the weapons produced there. The Springfield Armory continued to provide most US firearms until the middle of the twentieth century.

An uneasy compromise between a peacetime military and a perpetual warring machine meant that the government could never completely ease up on industrial intervention. Nor could it intervene in ways that were always equitable. The winners in this story were a cluster of arms contractors, a cohort of industrial capitalists, and a host of small manufacturers and inventors who occupied privileged positions in the geography and networks of production. White American civilians, too, benefited from ever cheaper fabrics and protection from belligerent European nations and from Indians who fought for their land. There were losers, too: manufacturers for whom government patronage was inaccessible or too uncertain; farmers who had little choice but to sell their land or water privileges; skilled artisans who saw mechanization replace their craft tradition; overworked laborers; and victims of American warfare. Amidst the costs and benefits of industrialization, federal officials, capitalists, and mechanics worked together, manipulated one another, and competed to create an industrial economy that would eventually rival that of a European nation-state.

In the closing decades of the eighteenth century, the American economy still existed as much to serve the needs of European markets as it did to secure US national security and self-sufficiency.[54] The United States could boast robust agricultural output and a thriving shipping industry and trade, but its markets were glutted with British manufactures, and it depended on Europe for military stores. Fast forward twenty or thirty years and it had achieved a military stalemate against one of the most powerful nations in the world, expanded its borders, become self-sufficient in arms production, developed a textile industry whose goods could compete with European fabrics, and stationed a corps of diplomatic agents in cities all over the world, ready to facilitate the sale of American manufactures overseas. It had begun with the recognition that independence required clothing and guns.

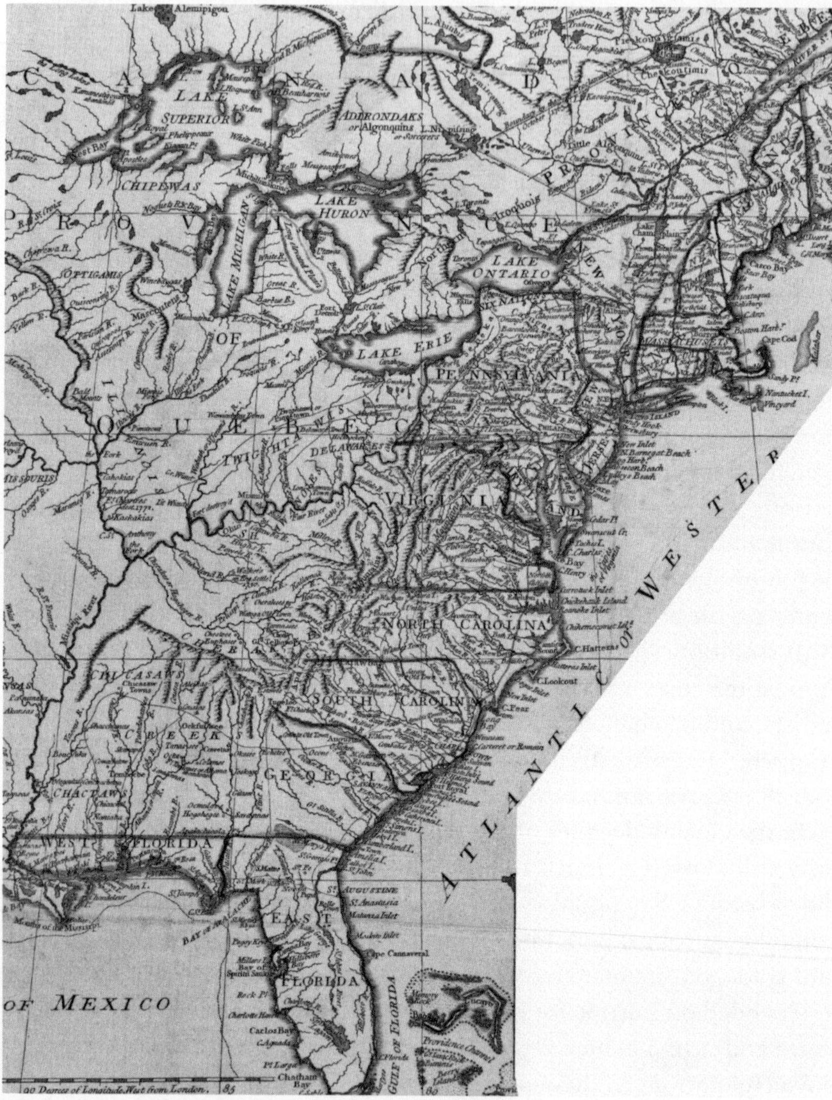

The American Colonies during the Revolution. From Robert Sayer and John Bennett, *The Theatre of War in North America* (London, 1776).
Courtesy, The Library Company of Philadelphia

[1]

"Our Naked Troops"

THE INFAMOUS WINTER at Valley Forge was just part of the story of suffering amidst shortages.[1] American rebels struggled to outfit their troops from the beginning of the war, when the Pennsylvania Committee of Safety begged for donations to clothe "our naked troops," until the ratification of peace.[2] The whole wartime experience was miserable for the soldiers who lacked supplies and for the officers who tried to provide them. Where and how, in a collection of colonies cut off from their supplier of manufactured goods, could one get blankets, weapons, even thread? The answer was: by whatever means possible. Eventually, this would mean from American manufacturers, but not in 1776, or even 1787. During the war, colonists scraped together most supplies from local and foreign merchants, from donations, and from stolen cargoes. George Washington's men were freezing, and as commander of the Continental Army, winning a war against one of the most formidable military powers on earth was not his only task; he also had to figure out how to outfit his troops.

The distress of American soldiers during the war is well known, as are Washington's lamentations about insufficient wartime supplies. These lamentations, however, are not known for their influence on American industrialization, even though Washington's writings on the necessity of military self-sufficiency shaped industrial policy up through the first years of the nineteenth century. Throughout the war, Washington wrote letters to Congress, state governments, and his military subordinates that stressed the consequences of insufficient arms and clothing supplies. In 1778, for example, he wrote to Congress, "The articles of clothing and blankets should also employ the utmost attention . . . : we are now in great want, particularly of the latter."[3] Three months later, however, the situation remained dire because "the state supplies of clothing hitherto sent to camp have been but small and partial."[4] Washington also wrote of the desirability of overalls instead of breeches for warmth in winter and coolness in summer, and he urged clothing agents to procure sufficient quantities. The army could not even mend the clothing it had "for want of thread."[5] Toward the end of the war, he continued to wait for clothing imports from France, and bemoaned the "miserable situation" soldiers faced, especially those from the South, who were "literally naked."[6] In 1781 Washington still desperately hoped that "matters of clothing will soon be put on a better foundation."[7] These experiences explain why, as president, Washington would commission Henry Knox to procure "superfine American Broadcloth" for his inaugural suit.[8]

When the Continental Congress declared Britain's North American colonies independent, it decided that Americans from then on would need to depend on themselves for protection. It thrust the colonies into a world of fiscal military states that knew how to provision an army.[9] Americans hoped for equal footing among Old World powers, but it would take decades of warfare and diplomacy before the United States confidently could supply itself with the stuff of war. Nonetheless, 1776 was a watershed year in the history of industrialization: it was then that Americans began to see manufacturing as more than a moral or social imperative. On the heels of patriotic boycotts and non-importation activities, local and Continental officials counted the domestic manufacture of war materials among their goals for the new nation. Their mission was to maximize the goods they could get from household manufacturing, urban workhouses, and small workshops and get those goods onto the backs and into the hands of soldiers. Inadequacies alerted them to the need for federal actions and a political economy that promoted economic diversification in the interest of self-rule. Wartime desperation and shortsighted pragmatism meant this would be a slow achievement, but in creating a Constitution to address the problems of independence,

Americans laid the foundation for the sort of civilian and martial manufacturing that would have prevented George Washington from wasting time writing about thread.

Colonial Precedents

Great Britain was America's point of reference for military administration and supply. Right after the Battles of Lexington and Concord, the Continental Congress adopted Britain's Articles of War.[10] Colonial presses churned out a series of British army drill manuals.[11] Many American military officers were bi- and trilingual and read many of the same military texts as their British counterparts. There existed a shared European martial culture, based on military etiquette and the "law of nations," of which some American officers strove to be a part.[12] At the same time, however, Americans prided themselves on their local militia. All of the colonies had strong militia traditions, except Pennsylvania (which did not have compulsory service until 1777).[13] These militias exhibited varying degrees of competency.[14] The most successful militia companies were usually from New England, where they were made up of men who knew each other well and operated from a sense of community. Militiamen from the thinly settled southern colonies, on the other hand, often barely knew one another, which made discipline and martial camaraderie difficult. Despite the relative skill of the Massachusetts Minutemen, following the Battles of Lexington and Concord and Bunker Hill it became obvious that some form of collectively organized army was necessary. Congress's solution was the Continental Army, a more competent and professional military body that required administration on a national scale beyond what they were accustomed to providing.[15]

In the British army, soldiers were given a .75 caliber Brown Bess musket with attachable bayonet and standardized uniforms that included a tricornered hat, shirt and stock, waistcoat, tight white breeches, gaiters, shoes, and a red outer coat with brass buttons. England had the manufacturing capabilities to make this sort of outfitting possible. Tens of thousands of households spun and wove the textiles for soldiers' clothing and blankets.[16] Over the preceding century, England had begun to manufacture its chief raw export—wool—which it had previously sent to Antwerp for processing. More and more, this manufacturing was done in factories that employed water frames (water-powered machines that spun cotton into yarn) and mechanized spinning machines.[17] Army colonels usually employed either a regimental agent or an army clothier to manage the fabrication, purchasing, and inspection of materials for uniforms.[18] Most arms were produced by gunsmiths in Birmingham, England. Britain, France, and Prussia had all developed robust arms-making capacities to provide weapons for their near-constant military endeavors. In England, arms manufacturing involved artisans who specialized in one part of the gun, like the barrel, as well as a gun maker who assembled the final product. These producers sold weapons to both private and public customers, but because the British government had a vested interest in keeping the industry humming in a way that suited its needs, it gave small government contracts during peacetime to supplement large wartime orders.[19] The Board of Ordnance closely scrutinized these firearms. Although the British government preferred to procure its military goods from a variety of dispersed producers for security and financial reasons, its decentralized system was highly regulated by officials who inspected and stamped privately manufactured goods.[20]

Once Americans no longer benefited from this regulation, they had to decide how much they would try to emulate it. On June 12, 1776, five days before the Battle of Bunker Hill, the Continental Congress created the Board of War to take charge of military administration. Although Britain served as a model in many ways, congressional delegates were wary of too much centralized power. They staffed the board with five members, all of whom were civilians. John Adams, as the board's first president, believed it was a mistake to rely on civilians to execute the war—a sentiment that stemmed in part from his resentment of the time military management took away from his other political duties, but also from his desire to measure up to the major military powers of Europe, who benefited from professional military administration.[21] It was almost impossible for the board to determine the quantity and quality of various militias' weapons and uniforms because there was no standardized system of supply. For the first several months of the war, officers turned in shoddy and incomplete returns. Supplies were low.[22] During the colonial era, militias were usually responsible for outfitting themselves, but when employed in British military service, American colonists often received red or blue coats, which some men still owned. Others had nothing.

The rebels faced the limitations of a colonial economy that existed to serve the needs of the mother country. Fortunately, they were not starting from scratch. Craft industries had existed from the earliest days of colonization and many Americans produced goods in their homes.[23] Britain subsidized the manufacture of naval stores and fuel and limited its own production of iron to stimulate colonial output. Iron ore, potash, copper, oil, flax, and hemp all counted among the items that received imperial bounties.[24] By 1775 the colonial iron industry produced thirty thousand tons—about one-seventh of the world's iron.[25] Several colonies also passed legislation to promote tanneries and the manufacture of cotton, woolen, and linens.[26] The metal trades flourished in Connecticut and Rhode Island, and papermaking, shoemaking, distilling, and shipbuilding had taken off throughout New England by the early 1770s.[27] One colonial manufacture that developed with minimal intervention from either local governments or the Crown was rifle making. German immigrants turned Lancaster and Berks Counties, Pennsylvania, into centers of rifle production starting around 1700.[28] Rifling technology had developed at the end of the fifteenth century in central Europe. Just as the hunters of German forests preferred the rifle's accuracy over the musket, so too did the American frontiersmen find the rifle perfectly suited to hunting and Indian warfare.[29] American gunsmiths made great improvements, and by 1750 their rifles had evolved into long, slender small-bore guns that even Europeans admired.[30] Once war broke out, the Continental Congress turned to Lancaster County for these reasons. New Englanders, on the other hand, did not need long-range rifles to hunt in hillier and more forested terrain; instead, they manufactured mostly fowlers (for hunting birds) and single-shot muskets.[31]

All of this production took the form of small-scale, highly skilled craft work rather than military industry. Colonial governments had to create incentives to transform the boycott energies of the 1760s into more substantial martial preparedness. Virginia offered a premium to the first person in the colony to produce five thousand pounds of gunpowder, and Massachusetts sent Paul Revere to Philadelphia to observe the best methods for making it.[32] Rhode Island, meanwhile, attempted to provide saltpeter works in every township. Maryland established its own gunlock factory in 1775, and the colonial congress of New York sought proposals for subsidized musket manufactories.[33] Much of what the states did was simply scramble to find supplies for their soldiers. Soon after the Battle of Bunker Hill, Massachusetts offered cash to any soldier who supplied his own blanket, and commissioned the purchase of stockings, shirts, and breeches. The New York Committee of Safety similarly authorized supply agents to purchase coarse woolens and blankets as cheaply as they could.[34] Pennsylvania's request for donations of coats and pants was thwarted by the navy's cutting up of "blankets, jackets, trousers, stockings, etc." to fix defective gun cartridges.[35] No state consistently met its pledge to supply two pairs of clothing for each of its soldiers, because, as Congress recognized, "the Manufactory of Woolens in these Colonies, tho' rapidly advancing, may not furnish an immediate Supply of Clothing."[36]

As president of the Board of War, Adams asked state assemblies to repeal any limits they had placed on the "Powers of their Delegates in this Congress," so that Congress could more readily "concert, direct and order" the war effort.[37] Congress's efforts to assume direct control over military administration, however, often exacerbated the problems of limited manufactures. It established the quartermaster general's office, the office of the commissary general, and the clothier general, but never clearly defined their duties, which contributed, for example, to a blanket shortage that ended up delaying bands of Continental troops from joining up with Washington.[38] When the Board of War had trouble procuring enough clothing for South Carolina troops, it entered into a problematic contract that ultimately led to a lengthy claims case against the federal government.[39] The national supply offices were not completely ineffective, but in general they struggled to outfit troops on a nationwide scale.[40] Even as the Board of War sacrificed fighting power for clothing—discharging able-bodied soldiers specifically so they could make clothing for the army—throughout the war, officers consistently cited insufficient clothing as the chief cause of sickness and desertion.[41] Its efforts to manage arms supplies were only slightly more successful. Because men who left military duty often absconded with their guns, the Board of War asked states to punish any soldiers who did not return their weapons when they left service.[42] After issuing calls "promoting and encouraging the making of good fire arms," it ordered the managers of the armory in Lancaster, Pennsylvania, and the gunlock factory in Trenton, New Jersey, "to deliver all the muskets and gun locks that each of them may have ready, in order to ensure the more expeditious arming of the continental battalion."[43]

These experiences represented efforts to centralize and militarize power, which was perhaps best symbolized by the marks on gun barrels that designated them property of the United States. Congress quickly realized that John Adams was right: civilians should not and could not handle the war effort. In December 1776, Congress granted George Washington wide-reaching powers as commander-in-chief, and the following autumn Congress authorized military officers to serve on the Board of War. General Horatio Gates replaced Adams as the board's president when Adams left for his diplomatic mission to France.[44]

Adams was much more comfortable as a diplomat than as a military administrator, but his approach to both roles exemplified security-based economic development in the new United States. Adams is not known for his theories on political economy. He did not adhere to any specific philosophy; his recommendations transcended coherent economic ideals. Just as Adams urged that military leadership replace civilian, he advocated for commercial arrangements that advanced military survival.[45] When Adams worked on the Model Treaty in 1776, his support for international commercialism bordered on economic liberalism, while after the war Adams worried about the dependencies that resulted from open trade, and eventually supported tariffs on manufactured goods.[46] During the war Adams pursued a pragmatic middle ground. For example, he encouraged a controversial trade deal with Russia to solve Americans' shortage of duck cloth for military supplies.[47] Americans would continue to import duck from St. Petersburg for decades.[48] Adams's stance embodied what would become the new nation's strategy of making real-time decisions in response to economic and military needs.

Realities of Independence

However much some Americans wanted to foster self-sufficient production while others never wanted domestic manufactures at all, merchants and officers desperately sought manufactured goods by whatever means possible, which often meant European imports and smuggled goods. American troops received much of their munitions from the smuggling network that connected France and the Netherlands to the North American coast via the Dutch West Indies. Additionally, the French government funneled military supplies to the Continental Army by loaning money to a French trading company that purchased arms at reduced prices from government arsenals and sold them to Americans on credit.[49]

The 1763 Charleville musket, the standard weapon of the French army, became the most common firearm used by American soldiers. The American rifle was more accurate than the musket, but it was also more difficult to load, and the labor intensity of its production rendered it impractical as the primary small arm. Additionally, many American officers resisted the weapon's departure from European norms. The musket, which delivered volleys of fire, was the preferred firearm for closed-rank eighteenth-century warfare where the target was not an individual but another line. Unlike rifles, muskets came fitted with a bayonet, which soldiers used when the lines closed. They also performed better in wet conditions, which could cause firearms to malfunction.[50] In many ways, rifles and sharpshooters embodied a distinctly American "democratic" mode of warfare in which any individual could aim with accuracy, a sharp break from the homogeneous infantry lines of European warfare.[51] But although Washington praised the accuracy of riflemen from Pennsylvania and Virginia, these soldiers were gradually phased out after the summer of 1776.[52] General Anthony Wayne said he would rather fight with a bayonet and no ammunition than with a rifle.[53] When General Baron von Steuben took over training the Continental troops in 1778, European prejudices prevailed and the French Charleville musket remained the standard for years after the war, providing the blueprint for the US Model 1816 musket.[54]

Uniforms, too, came from France. Americans started receiving French blankets and suits of uniforms in 1778. Clothing Department regulations the following year acknowledged that clothing undoubtedly had to be imported.[55] Clothing agents also got uniforms and other goods from prize cargoes, but there were large gaps in shipments, and agents spent a lot of time waiting for goods to arrive.[56]

Perhaps because it all worked out, Americans forgot. Following the resumption of peace in 1783, wartime struggles receded from political memory as such pressing issues as state and national debt occupied policy makers' time. When manufacturers petitioned the Confederation Congress for support, their pleas often fell on deaf ears.[57] When, for example, a congressional committee read the memorial of steel manufacturer Stacy Potts, it determined that, "considering that there are at this time many important manufactures in the different States . . . large sums of money [can] be saved which must otherwise be expended in foreign Countries, circumstances which deserve the most serious consideration at a Period when the United States are labouring under heavy debts both foreign and domestic." The committee's solution was to suggest that state legislatures "countenance and encourage the establishment of useful manufactures."[58] Just the previous year, Congress had still been trying to enforce prohibitions of British imports, but legislators' refusal to support manufacturers once the peace process was underway signaled a shift in priorities. Indeed, they must not have been too bothered by British minister David Hartley's argument that "the introduction of American manufactures into Great Britain can be of no service to either."[59] During treaty negotiations, Hartley proposed that American commodities like tobacco be imported into Britain, and British manufactured goods into the United States, without duties. The post-independence trade relationship caused enough distress among manufactures to prompt state lawmakers in New England and Pennsylvania to pass tariff laws in the mid-1780s, but these were local, not national.[60]

Under the Articles of Confederation only states could levy taxes, which made national protectionism an impossibility.[61] Of more immediate consequence was the nation's inability to implement a tax policy to pay off its wartime debt, which threatened to impede economic independence.[62] When Congress requested tax revenue from states, it received mixed responses. As James Madison pointed out, "a distrust of the voluntary compliance of each other may prevent the compliance of any."[63] New York and Rhode Island flat-out refused to cooperate with a tariff. States that did crack down on tax collection faced armed protests like Shays' Rebellion in western Massachusetts, which signaled to potential investors that fiscal accountability was a problem. If the United States were to become a peer of Britain and Europe, this economic and military insecurity could not continue.

Constitutional Foundations for Manufacturing

When constitutional delegates met in 1787 to modify the structures and powers of the national government, they were much more concerned about paying down Revolutionary War debt and attracting foreign investors than they were about manufactures, and their major debates centered on democracy, representation, and slavery. Legal scholars, political theorists, and historians have interpreted the document as a treatise on economic interest or Enlightenment political ideology, as pro- or anti-slavery, and as a mandate on strong or limited federal power.[64] It is certainly not an explicitly pro-manufacturing doctrine, especially because support for the Constitution did not track economic or regional interests. In fact, manufacturers did not necessarily support the Constitution, despite its association with Alexander Hamilton's nationalist economic agenda. Even strong proponents of manufacturing protectionism, like Bostonian Benjamin Austin Jr., for example, opposed the Constitution because it would create a federal government that was too expensive and powerful but would not actually compel Europe to end the trade restrictions that hindered American commerce. Austin preferred to revise the Articles of Confederation to allow for national custom duties and other commercial legislation, rather than adopt a new constitution.[65] Ratifying conventions in states with burgeoning industrial production, like Massachusetts and Pennsylvania, divided over support for the Constitution.[66]

Even if we cannot equate manufactures' interests with constitutional support, the new Constitution provided improved institutional scaffolding for manufacturing. Under the Articles, not only could the government not tax; it could not grant inventors privileges that the states were bound to respect, for which manufacturers also petitioned the government.[67] Article One, Section Eight, of the US Constitution remedied both of these shortcomings by giving Congress the power to levy taxes and grant exclusive rights to inventors in the form of patent and copyright protections. Policy makers generally viewed monopolies as anathema to the American experiment, but many also believed that property rights were necessary to encourage intellectual and mechanical innovation. Although the exact intentions of the framers are unclear, the drafters most likely included the phrase "to promote the progress" in order to clarify that these rights were supposed to have a broader purpose than individual profit.[68]

The Constitution's inclusion of monopoly-granting powers was not its only seemingly abhorrent Old World feature. The framers included in the Constitution the right of the federal government to raise and direct armed forces, even as Americans feared the existence of centralized military power.[69] The violent reality of tax uprisings, like those led the previous year by Daniel Shays in western Massachusetts, made the Constitution's proponents aware that "a certain portion of military force is absolutely necessary in large communities."[70] And so, even as the regular army was reduced to barely one thousand soldiers, the framers created the mechanisms by which the government could grow it.

While much of Article One, Section Eight—especially the clauses related to tariffs, patents, commerce, and military funds—would help shape the course of industrial development in the new nation, it was the Constitution's provisions for creating and staffing departments within the executive branch that would ultimately prove transformative. Although the delegates largely responsible for explicating executive power—Alexander Hamilton, James Wilson, and Gouverneur Morris—advocated a strong executive, the majority did not set out to create one. During their five months in Philadelphia, delegates took up the issue of the executive branch several times. They agreed to a single executive on June 4, but did not work out the details of its election or powers until September.[71] As late as September 8, they continued to debate whether executive power should be dependent on the legislature for actions like treaty making.[72] During the penultimate day of the Convention, they agreed that the president would be able to appoint major officers, such as ambassadors, and that he would not need the Senate's advice or consent for lower-level officers. Delegates had defeated the motion for a council of advisers, but Article Two, Section Two, Clause One, ultimately gave birth to the presidential cabinet by giving the president the power to "require the opinion of the principal Officer in each of the executive Departments." The fact that there were appointed cabinets whose members could act without mandate from Congress meant that they would be able to bolster the nation's military and its manufacturing sector without waiting for legislation.

When constitutional delegates met in 1787, domestic manufacturing and military self-sufficiency were not their primary concerns, but they built them into the fabric of the Constitution. The document they created was ultimately about security. It empowered federal officials to foster industrial initiatives so that generals did not have to worry about thread or distribute smuggled arms to their troops. In the midst of historical amnesia about wartime hardship, executive officials had the power to prevent another winter like Valley Forge from happening again.

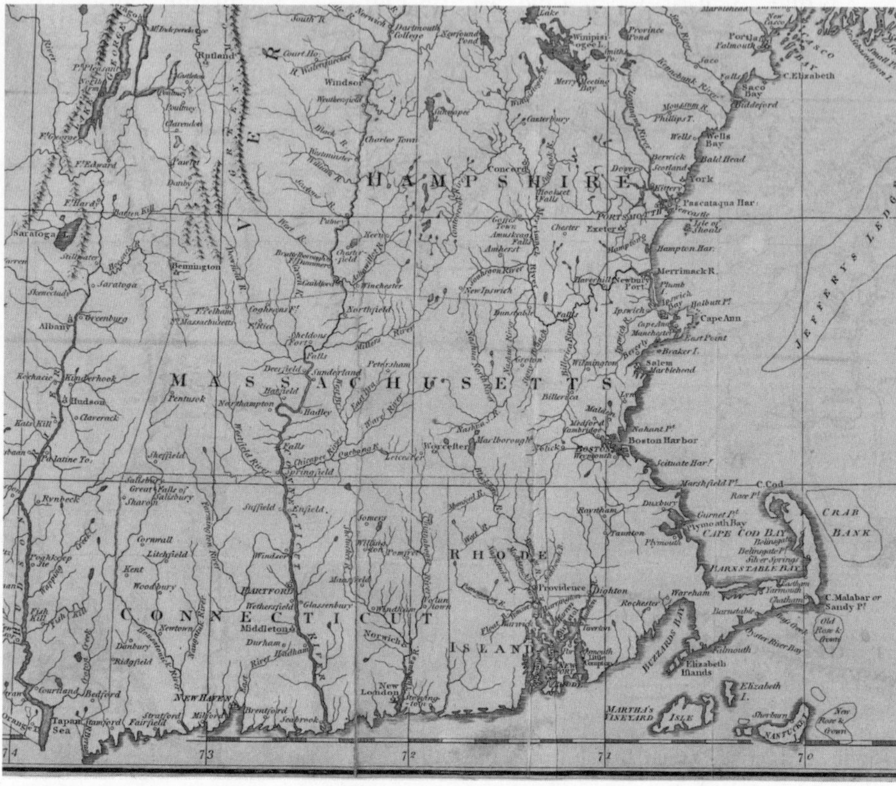

New England, 1790s. From John Russell, *Map of the Northern, or New England States of America: Comprehending Vermont, New Hampshire, District of Main, Massachusetts, Rhode-Island, and Connecticut* (London, 1795). Courtesy, The Library Company of Philadelphia

[2]

The Political Economy of Guns and Textiles

TENCH COXE WAS INCENSED. The former delegate for Pennsylvania to the Continental Congress had read *Observations on the Commerce of the American States*, which "disseminated ideas, very unfavorable to the United States" about its capacity for industry. Written by British politician John Baker Holroyd, Earl of Sheffield, the economic tract denigrated American commerce and industrial prospects and advocated the maintenance of restrictive commercial policies against Britain's former colonies. Coxe, aware that Sheffield's writing was extremely influential among members of Parliament, published a rebuttal. American manufactures, Coxe asserted, "were found to be the most successful competitors with those of Great Britain."[1] Most European travelers and political economists, however, disagreed with him.[2]

Following their war for independence, Americans had pushed aside concerns about economic patriotism and military supplies and backed off their manufacturing initiatives. They sought political and military independence but struggled to thrive outside their privileged postcolonial relationship with Britain.[3] Many feared that industrialization and militarization would threaten their experiment in republican government, and they viewed the European industrial city and standing army with trepidation, even horror. Anxieties aside, economic independence and national defense were powerful motivators for the newly independent US government. As the Old World watched the United States, Americans jealously guarded their economic reputation. War against European nations could happen again at any time, and warfare against Native Americans was constant. The American nation-state, as well as its economy, needed protection. How, though, would that protection be achieved?

Federal officials debated how best to create a sound economy and a secure nation-state in the face of commercial pressures at home and abroad. These early debates about industrialization yielded mixed results, but within twenty years of its founding the United States would have its first successful mechanized cotton mill and a military establishment that included not only state militias but a professional army, a navy, and a national military academy.[4] America's military-industrial beginnings, almost imperceptible at the time, had their roots in federal decision making that occurred in the 1790s in the shadows of European precedent.[5]

Decisions

The US Congress assembled for the first time under the new Constitution on March 4, 1789, in New York. On April 30, several weeks after the Senate and House of Representatives finally achieved a quorum, George Washington was sworn in as president. Federal legislators had many things to deal with besides manufacturing, but what they decided laid the groundwork for its development. As Congress tackled public debt, commerce, and the maintenance of armed forces, legislators set the stage for the executive department to exert autonomy over industrialization.

One of Congress's first tasks was to amend the Constitution. The provision that became the Second Amendment established the citizen's right to keep arms for the protection of himself, his state, and his nation. Although this amendment quickly faded from national consciousness until the twentieth century, its passage had some immediate implications for manufacturing. America's tradition of local, "well-regulated" militias had begun to break down during the Revolution because of desertions and disappointing performances in battle. Following the war, Federalist desires to use the militia to put down tax uprisings bumped up against Anti-Federalist fears of federal control over state militias.[6] In this context, the passage of the Second Amendment was a rebuttal to insults or attacks on the militia, and a national affirmation of the armed citizen as a bulwark against tyranny. In conjunction with provisions in the Constitution for arming the militia, however, the amendment can be understood as an acknowledgment of the importance of the federal government, which would not only guarantee the right to have a gun; it would physically provide one. American citizens did not, in fact, produce enough affordable firearms to arm themselves, which was why their exportation was prohibited and their importation tax-free during the early 1790s.[7] As a result of its constitutional obligations, the federal government spent the 1790s and early 1800s determining how best to shore up American firearms manufacturing.

A more general discussion of domestic manufactures first appeared on the new federal government's agenda during decisions about taxation and the national debt. Lawmakers agreed that a tariff was the best way to raise money, but they debated what specifically should be taxed and how much.[8] The resulting legislation levied moderate duties on spirits, wine, and coffee, as well as on several of America's more established manufactures. Tallow candles, for example, were taxed at two cents per pound, while wool and cotton cards were fifty cents per dozen, and nails and spikes were one cent per pound. Yarn was included at ninety cents per 112 pounds, but notably absent from the list of specific duties were firearms and textiles. At this stage, Americans did not manufacture enough of either to warrant interferences with their importation. The bill passed easily and was signed into law on July 4, only several months after Washington's inauguration.[9] It put into practice the federal government's power to tax, and it illustrated the importance of domestic manufactures, but it was not substantial enough to result in any real boost to industry. Over the next few years, as Congress needed greater revenues to tackle the Revolutionary War debt, it expanded its list of taxed goods. But even then, wool and cotton cloth were taxed at some of the lowest ad valorem rates, while firearms were duty-free in 1794 and 1795 and from 1797 to 1801.[10]

Of more assistance to domestic manufacturing was the creation of executive departments to oversee war, diplomacy, and finance. Congress created the Department of Foreign Affairs on July 27 (to be renamed the Department of State two months later), the Department of War on August 7, and the Department of Treasury on September 2. The Department of War would become the biggest champion of domestic industry. Within five months of its establishment, for example, it forwarded a report to George Washington on the organization of the militia, in which Secretary of War Henry Knox asserted that it was "unworthy of the dignity of a rising empire to depend on foreign supplies for its defense." In addition to arms and artillery, "clothing for the troops," he said, "could with ease be manufactured within the United States, and the establishment in that respect would tend to the encouragement of important manufactories."[11] The Department of State was less vocal about domestic manufacturing, but its tasks of corresponding and negotiating with other countries' ministers, overseeing US consuls, and responding to foreign memorials would determine American manufacturers' ability to profit overseas.[12] Additionally, in the year following the creation of the first executive departments, Congress decided to house the patent office within the Department of State.

Congress had barely been in session when mechanics began soliciting their legislators for monopolies.[13] Patents had a long history in America; the colony of Massachusetts had issued the first patent for a mechanical invention in 1646 for improved sawmills and scythes.[14] Until 1789, patent conferral happened at the state and local levels, but the Constitution's Article One, Section Eight, Clause Eight, gave Congress the power to grant exclusive rights to inventors. The frame of reference for national patent legislation was Britain. Inventor James Rumsey, for example, commented that there "was a Committee of Congress appointed to bring in a bill for Establishing an office for granting Exclusive Wrights to inventors &c. . . . England I beleive [*sic*] has fixed it on the best Establishment."[15] Thomas Jefferson tried to counter Americans' proclivity to emulation by arguing that the American patent system should use "reason for our guide, instead of English precedent."[16] Interestingly, although Jefferson opposed monopoly rights for machines or processes that might otherwise be used freely to the benefit of society, in this 1789 letter to James Madison he recommended a patent term of nineteen years, which was five years longer than the British system (and the same as Jefferson's time frame for how often the Constitution should be amended).[17] Congress chose emulation. The Patent Act of April 10, 1790, delegated the responsibility of patent assignment to the secretary of state, the secretary of war, and the attorney general, who could grant monopoly rights for a "useful manufacture, engine, machine, or device, or improvement therein not before known or used" for up to fourteen years.[18] Mixed feelings about monopolies aside, the new patent legislation was more equitable than Britain's because the US application process was relatively easy and inexpensive, while fees for a British patent that covered England, Scotland, and Wales were more than ten times greater than per capita income.[19] British inventors were also less likely to apply for patents because the state could appropriate patent rights without consent or compensation.[20] Prizes and premiums were more common, but these were often reserved for the elite.[21] Private parties in the United States offered premiums for discoveries and inventions as they did in Great Britain—Samuel Slater, in fact, had been enticed to emigrate by a newspaper report about premiums awarded by the Pennsylvania Society—but Americans, by and large, would seek patent rights more than awards or prizes.[22] The American Academy of Arts and Sciences, for example, offered a biennial award, but could not present one until 1828, for want of candidates.[23]

Congress briefly debated subsidizing the immigration of skilled workers from Europe, as a sort of technological piracy, but settled on the patent system to incentivize mechanical innovation, in conjunction with a lax immigration policy. On March 26, 1790, it passed an "act to establish an uniform rule of naturalization," which provided citizenship to free white immigrants of "good character" after only two years of residence.[24] This policy stemmed from attempts during and after the war, from both private business agents and such political leaders as Thomas Jefferson and John Adams, to lure skilled workers from abroad.[25] Recognizing that international competitiveness rested on the immigrants the nation could attract, Benjamin Rush published "Information to Europeans who are disposed to migrate to the United States" to advertise the new immigration policy and entice "mechanics and manufacturers of every description, [who] will find certain encouragement in the United States."[26] Rush boasted that "we are already nearly independent of the whole world of iron-work, paper, and malt liquors; and great progress has been made in the manufactories of glass, pot-ash, and cloths of all kinds. The commercial habits of our citizens have as yet prevented their employing large capitals in those manufactories; but I am persuaded that if a few European adventurers would embark in them with capitals equal to the demand for those manufactures, they would soon find an immense profit in their speculations."[27] Rush perhaps oversold the status of American industry and the opportunities it provided; nevertheless, for potential emigrants the United States offered an opportunity beyond what Britain's middling mechanics could hope to achieve.

The Formative Decade

In 1794 Henry Wansey, like other curious visitors to the United States, set out to "mark its progress since it began to rank among the nations of the earth." As a British textile entrepreneur, he was interested in American manufactories, but he described those he saw as nothing more than "seeds of manufacturing." He assured his readers that England need not "fear a rivalship there." Wansey's major conclusion was that American manufacturing would never pose a threat to British industry, nor even keep pace with population growth. Instead, US growth and prosperity would be a boon to British manufacturers because Americans, already dependent on British goods, would only increase their consumption of England's manufactures as their population expanded. In his estimation, it would be "at least a century" before the United States could manufacture for its own population.[28]

Wansey's ruminations were at least partly right. The first Congress had laid the groundwork for industrial development, but the United States was still very much reliant on British trade and manufactured goods; indeed, its commercial relationship with Britain largely had resumed its prewar course.[29] This was a cause for concern among some policy makers, who, like John Adams, worried that unless the United States developed more manufactures, it would exist perpetually as a colonial outpost of Britain.[30] Wansey observed that the United States seemed to have "*many* better sources of national wealth, at present, than manufacture," but this was only because the federal government had not yet committed to industrial development.[31] In 1794 policy makers still wondered: Should the United States export mostly agricultural produce and to whom should it sell? Should it continue to import manufactures from Europe or develop its own industry? How large should that industry be?

Something was happening, though, that made possible what seemed to a later observer "contrary to the natural course of things."[32] Wansey's tour, in fact, occurred just as the federal government decided to establish national armories, and soon after the wheels on America's first successful mechanized cotton-spinning mill in Pawtucket, Rhode Island, started turning. Policy makers were figuring how best to manage America's diplomatic and commercial position vis-à-vis the rest of the world, while aspiring manufacturers pooled resources to open mills. Over the ensuing decade, direct investment for military production created a national arsenal of domestic weapons. Most manufacturers, however, were left to navigate individual state aid (typically subsidies, lotteries, and local tax breaks), private wealth investment, and the recruitment of foreign talent. Results were uneven, but the Industrial Revolution had begun.[33]

Following independence, there were two major strands of thought about how best to secure America's place in the world via its economy. Alexander Hamilton was one of the staunchest advocates of an industrial society propped up by a strong government and wealthy capitalists. He issued *A Report on Manufactures* in 1791 to recommend policies the United States could follow to elevate its economic status above that of a raw-goods producer. Hamilton argued that the United States suffered from an inferior position in the international system of commercial exchange that rendered the nation vulnerable to both internal and external threats. The solution, according to Hamilton and like-minded Federalists, was domestic manufacturing.[34] To overcome the "scarcity of hand, dearness of labour, and want of capital" that threatened to prevent the United States from developing a robust manufacturing sector, Hamilton proposed a policy of bounties and premiums to stimulate investment and production.[35] The government could, he said, create demand with its economic policies. While James Madison argued that higher duties on foreign manufactures would free the United States from economic dependence on England, Hamilton advocated a moderate impost that would be collected in ports and paid back in bounties to reward certain economic activity.[36] Hamilton's mercantile and British sympathies prevented him from favoring burdensome tariffs, but like most Federalists, he believed federal stimulus coupled with the national development of a strong militia was the nation's best protection of its independence.[37]

Thomas Jefferson, on the other hand, championed a national economy propelled by virtuous yeoman farmers.[38] He believed a strong military establishment and large-scale manufacturing threatened, rather than secured, independence. For Jeffersonians, the ideal economy for a republican society was one based on highly developed agriculture and the ability to dispose of produce in international markets.[39] This idealized image, however, was an impractical one: it could not exist in a world of imperial rivalries, open borders, and resource insecurities. Even if, as Jefferson urged, international markets absorbed American produce, these markets would not be enough to preserve independence, nor would they enable a nation of yeoman farmers to withstand geopolitical pressures.

Additionally, while many Jeffersonians staunchly opposed the expansion of naval and military capabilities, Jefferson greatly admired French arms making, and during his tenure as minister to France wrote home to fellow federal officials about the virtues of arms manufacturing.[40] Textile manufacturing, too, had its virtues. The Jeffersonian political economy depended on agriculture, but its full realization also included the manufacture of coarse cloths and other homespun goods.[41] The Revolutionary experience had illustrated to Jeffersonians that these types of manufacturers were part of the republican vision so long as American wares were "useful," in contrast to the fripperies produced in European factories.[42] Small-scale manufacturing, in fact, separated the developed agrarian society from the undeveloped one. It was the natural extension of a successful agrarian republic and required no corrupting government intervention. Any manufacturing that did develop should do so naturally, but because Jeffersonian political economy depended on foreign consumers, federal action sometimes became necessary to maintain commercial exchange. For example, as president, Jefferson decided to wage war against the Barbary States in the early 1800s, rather than pay tribute to maintain American trade in the Mediterranean.[43] Opposition to naval buildup and government intervention often fell apart in the face of threats to commerce.[44] The impracticality of the Jeffersonian persuasion—of a nation that depended on the individual farmer for its independence—is most evident in the decisions that federal officials and private entrepreneurs made in the decades following the Revolution.

Neither Hamilton's nor Jefferson's vision reflected what happened on the ground. Many manufacturers wanted government intervention, but their positions on the economy did not necessarily track Hamiltonian or Jeffersonian ideals. Following the Revolution, mechanics saw themselves as vital to national security; they had produced goods for a colonial economy cut off from its industrial supplier, and their wares provided material sustenance for independence. Once the war ended, however, British goods glutted American markets and mechanics wondered what the nation's economic fate would be. They channeled their economic usefulness into a political identity and ideology, and they advocated for a production-centered economy.[45] In the years immediately following independence, many artisans and mechanics supported Federalist policies that championed an activist state. They entered an uneasy alliance with merchants in order to advocate government regulation of trade, but as Federalist policies began to privilege wealthy mercantile interests over their own, manufacturers increasingly turned toward Jeffersonian policies.[46] After 1800 they sought economic equality from the Democratic-Republican platform, which promised to democratize not only politics but banking opportunities as well. A dramatic increase in the number of state-chartered banks equalized access to banking and credit, which enabled small producers to experience greater financial autonomy, seeing themselves as the urban versions of Jefferson's yeoman farmers. Even as they rejected the tenets of a Hamiltonian system of development whose oppressive state seemed to foster inequality, however, most manufacturers still wanted government-sponsored internal improvements and protective tariffs.[47] Regardless of which political ticket manufacturers voted, many desired an activist state that intervened in the economy to protect American manufactures.

Manufacturers of all stripes petitioned Congress for various forms of protection and support, but with mixed results. Just as the 1789 tariff was proof of the importance of manufacturing to the national agenda, so was the 1794 tariff, which offered support for boots, shoes, and slippers, millinery, iron manufactures, stones and marble, carpets, leather, carriages, paper, medicines, and cotton and linen manufactures, in addition to gold, silver, tobacco, and a variety of foodstuffs.[48] Cordage (rope) manufacturers requested and received support, despite their relative success without protection. Because the utility of ropes, especially for shipping, was not a hard sell, the Committee on Commerce and Manufactures supported a drawback for exporters of cordage, lines, twine, and packthread manufactured within the United States.[49] Congress even granted the request of John F. Amelung, a glass manufacturer from Frederick County, Maryland, who had petitioned for a loan from the government because of financial losses he had experienced from his manufactory. Amelung's manufactory was deemed "of great consequence and of great utility to the union," and Congress authorized the secretary of treasury to make him a loan of $8,000.[50]

There were limits, though, to what the federal government would do for its manufacturers. Seven years after Congress granted Amelung his loan, William Crowley Jordan of Philadelphia requested public capital from Congress to help him establish a silk manufactory, but he was subsequently denied. That same year, the proprietors of a glass manufactory in Boston, as well as a group of "sundry manufacturers of hats," petitioned the Committee of Commerce and Manufactures for additional duties, which were, in the words of the committee, "of course disagreed to." The committee maintained that the real problem was the high cost of labor, which "time only could cure."[51] After initial duties were established, Congress was disinclined to increase tariffs in favor of manufactures, not because manufactures were unimportant to economic security but because support for them had to be balanced among a host of economic factors.[52]

The re-export trade made up half of the United States' commerce, and the nation still had to consider its relationship with Britain. The values varied each year, but Britain generally absorbed around 20 to 25 percent of America's produce following independence, and the provisioning of Britain's colonies in the West Indies was particularly important for American merchants and suppliers of foodstuffs and naval stores.[53] The French Revolutionary Wars, however, threatened the safety of US commercial voyages, while British wartime trade regulations challenged their legality with an expanded definition of "contraband," meant to justify the seizure of France-bound ships. At issue was the sanctity of the United States' status as a neutral provider. American policy makers recognized that diplomatic negotiations with Britain were of paramount importance but disagreed over the substance of these negotiations. James Madison, serving Virginia in the House of Representatives, believed that the United States should inflict commercial hardship on Britain. His argument that Britain needed American raw materials and markets, while the United States could easily do without British manufactures, rested more on moral sentiment than on economic reality. He and other Jeffersonian Republicans advocated a complete redirection of American trade from Britain to France. Federalist merchants and politicians, on the other hand, were loath to give up trade with Britain and instead requested defense measures that included raising an army of twenty thousand men in order to convince Britain to change its policies. Neither plan came to pass. The treaty brokered in 1794 by diplomat John Jay avoided war with Britain and offered Britain most-favored-nation status. While it offended many Americans' "national self-esteem," according to historians Stanley Elkins and Eric McKitrick, its ratification marked a diplomatic success for a new nation struggling to improve its international commercial viability.[54] The treaty secured Britain's promise to compensate American merchants for maritime losses and, although the United States agreed to adhere to Britain's anti-French blockades, US maritime commerce improved as a result of Jay's negotiations.[55]

The Jay Treaty brought to the fore the issue of how best to determine and negotiate the composition of the United States' commercial offerings, as well as their recipients. It did not, however, resolve it. When Jay's Treaty was ratified by Congress in 1796, Britain earned most-favored-nation status, which meant that the United States could not levy excessively burdensome duties on British imports.[56] Tariffs would not and could not be the government's primary means of industrial encouragement. Instead, policy makers would encourage technological improvement via the patent system and the immigration of foreign mechanics and importation of foreign machines, relying on rhetoric and state aid to do the rest. It was clear that the federal government valued manufacturing, but domestic and international considerations left military production the sole recipient of significant federal support. Mixed messages made a rocky start to American industrialization. As we follow Wansey's tour of New England, we will see what manufacturing looked like on the ground.

Clothing for Citizens

The textile industry struggled to take off in the face of policy inconsistency. Even as public figures extolled the virtues of American clothing, they failed to rally around a cohesive industrial plan. While policy makers encouraged immigration, granted monopolies for inventions, and permitted technological piracy, they largely left civilian manufacturers to their own devices. The first substantive tariff would not be raised until 1816, and individual state subsidies failed to make textile manufacturing profitable for all but a handful of manufacturers.[57] In the interim, textile factories opened and closed.[58] There were, though, some moderate success stories. Both Massachusetts' Beverly Cotton Manufactory, which employed more spinning machines than any other firm in the country, and Rhode Island's Almy, Brown, and Slater, which experimented successfully with the water frame, have been credited as the first successful textile mill in the United States. Only the latter, though, survived beyond the first decade of the nineteenth century.[59] Both undertakings were backed by wealthy merchants, but the Rhode Island firm was less ambitious, employing fewer machines and producing mostly cotton yarn and thread for household weavers and merchants who coordinated outwork. It also benefited from loose US immigration policies and the encouragement of technological piracy, the first meaningful steps toward a national industrial policy. In general, though, inadequate state and federal policies prevented textile manufacturers from capturing the domestic market until commercial warfare against Britain in 1806 halted foreign imports.

It was not that textiles were unimportant. Even if military efficacy did not depend on textiles, the republican experiment did. Commercial dependence of all kinds was anathema to republican success. Textiles—yarn, thread, broadcloths, and eventually ready-made clothing—were visible forms of dependence. President George Washington recognized this fact when he addressed Congress wearing Connecticut-made broadcloth.[60] As historian Michael Zakim has shown, textile production was intrinsically connected with national ideas about the productive citizen. Americans' ability to manufacture the cotton, wool, and linen for clothing made evident the virtuosity of their struggle for independence.[61] A national ideology that supported textile production—first at home and then in the factory—helped reconcile the transition from republican agrarianism to industrial nation-state. Just as the armed citizen who was ready to protect his nation was imbued with a republican morality, so too was the citizen who, if she did not make her own, purchased American-made thread, yarn, and cloth.

Yet, morality was one thing, reality another. Although the nation's first tariffs offered some protection to manufactures, they paled in comparison to tariffs forty years later and did little to help budding industry.[62] The 1790s created a sort of policy conundrum in terms of domestic manufactures: Americans sought economic independence, but their hands were tied by Jay's Treaty and by merchants' reliance on foreign markets.

Policy uncertainty was in fact why so much of what Wansey observed were "seeds of manufacturing." On his first stop, in the city of Boston, Wansey greatly admired the bridges and theaters there but dismissed all the manufactories he saw, including the "curious" wool-card manufactory that produced the cards—devices that disentangle and clean raw wool—used by "every housewife" in the area.[63] Boston industry, however, was not quite as feeble as Wansey observed. He perhaps did not realize that he was in the city known for early industrialist Paul Revere, whose experiments with copper rolling contributed to the shipbuilding industry. Shipbuilding was one of Boston's most prominent industries, and its decline in the years prior to the Revolution—a result of high labor costs—had started to reverse as a result of federal naval demands. Subsidiary industries, too, like sailcloth manufacturing, were on the upswing, after struggles in the 1780s to earn profits beyond what the state government provided in bounties.[64]

Wansey's disdain continued when he left Boston. Heading northwest out of the city, he stopped in the "straggling village" of Waltham.[65] Other than the production of some very rough homespun kersey—a cheap, coarse woolen cloth of twill weave—nothing belied the fact that the town would become a textile center and certainly not that it would become home to the nation's first integrated textile factory system twenty years later. Historians have tended to characterize the success of the Waltham-Lowell System—the nation's first large-scale, integrated production process—in the 1810s and 1820s as the result of innovative businessmen and large sums of merchant capital. Yet in 1789 some of the same men associated with Waltham invested their wealth from trade in textiles and failed. The difference? In 1789, the Massachusetts Legislature granted a group of wealthy merchants, including Israel Thorndike, George Cabot, and Joseph Lee, a parcel of land and the incorporation of their textile manufactory. Following incorporation, the state agreed to provide $4,000 in aid, so long as the company produced within seven years at least fifty thousand yards of cotton and linen cloth that had theretofore been imported. In 1798 the original proprietors sold the factory, and within ten years the company ceased all operations.[66] Almost twenty years later, however, the Boston Manufacturing Company paid its first dividend of $170 per share, and Israel Thorndike and his son, Israel Thorndike Jr., one of the company's directors, each made $3,400. The Boston Manufacturing Company had benefited from wartime consumer demand and a postwar tariff tailored to their business needs.[67] These dividend payouts increased and were issued twice annually, as the company grew and received an influx of federal capital in the early 1820s.[68] The disparity in outcomes in the 1790s and the 1820s reflected the scale of aid. In 1790, when the stakeholders of Beverly Cotton Manufactory, who received state aid to produce coarse cloth, sought federal patronage and an additional impost on cotton imports, they did not receive it.[69]

As Wansey toured New England, he saw the limits of individual state aid. From Waltham, he traveled west and stopped in Worcester to dine on beef and veal down the street from what had been the Worcester Cotton Manufactory. Three years earlier its promoters announced in a Massachusetts newspaper that the factory had produced its first piece of corduroy. Eighteen of Worcester's leading men invested in the mill, which was built on a small stream parallel to Main Street. Account records indicate that they employed local men and women in weaving the company's menswear fabrics and paid city residents for boarding female workers. James Blake, for example, received compensation for picking cotton, making candlesticks and loom irons (instruments that press fabric in the weaving process), and "boarding sundry girls."[70] The proprietors of the company were optimistic, extolling the superiority of their fabrics over imports and purchasing a large amount of pine boards in January of 1790, indicating an expansion of factory size.[71] Yet, a year later, hundreds of yards of fustians (linen/cotton textiles) and jeans were returned to the company unsold, and most of the shareholders divested themselves of interest. At the time business ceased, there remained £288 worth of fustians and jeans and £37 worth of linen yarn, which were distributed among shareholders.[72] Nathan Patch, the sole remaining proprietor, petitioned the Massachusetts legislature for aid in continuing the business, arguing that "an encouragement to American Manufactures is of great importance to the welfare of this community" and that his manufactory would "preclude the necessity of many Importations."[73] On February 20, 1792, Patch was granted tax exemption for ten years, but that status mattered little when the factory was sold and converted into a store soon after.

Alexander Hamilton's celebration of the "publick spirit" of Massachusetts in regard to manufacturing belied the fact that the failures of the Beverly Cotton Manufactory and the Worcester Cotton Manufactory were not uncommon.[74] State tax breaks did little to make cloth production profitable. When Wansey traveled south to Hartford, Connecticut, he found a woolen manufactory, established six years prior, "very much in decay."[75] This was the factory of Jeremiah Wadsworth, who had provided the cloth for Washington's Congressional-speech suit. Wadsworth advertised the high quality of his mill's products, marketing its goods at the same prices as British ones. Consumers apparently disagreed, and the company disposed its goods at auction and shut down within several years.[76] Pretensions to high quality aside, the market was inhospitable to American-made cloths, and state governments could do little to remedy this. In fact, the limits of state governments were painfully apparent to the small start-ups Wansey observed throughout New England.[77] State governments were certainly important for economic growth—they had the power to grant corporate charters, oversee internal improvements, issue lotteries to raise money for new business ventures, and award tax exemptions on company profits.[78] Only Congress, however, had the authority to regulate trade among states and with foreign nations, which was necessary to sell American-made goods profitably in domestic markets.[79] In the words of Elisha Colt, the manager of the Hartford Woolen Factory, "the Legislature of the United States alone are competent to this business [of encouragement]—the separate States having neither the authority or funds for the purpose."[80] Like the Worcester Cotton Manufactory, the Hartford Woolen Factory had received state aid: "a trifling bounty the first year on spinning," as well as a two-year exemption from a poll tax on their workmen and workshops. Textile factories in New Haven, Farmington, and Killingly, Connecticut, received the same benefits, but these were short-lived. The state legislature had also issued the Hartford Company a lottery to raise £1,000 for procuring new machinery to extend the factory's business. The directors correctly predicted that the $3,000 they would raise still would not be enough because their want of capital was so great.[81]

One early success story, however, had escaped the attention of Wansey. Forty miles down the Blackstone Valley from the Worcester Cotton Manufactory, just as the company paid out its shareholders in leftover linen yarn and "goods on hand," the waterwheel on Samuel Slater's mill first started turning.[82] Often heralded as the birthplace of the American

Industrial Revolution, and Samuel Slater as the father of the factory system born there, the Rhode Island firm of Almy, Brown, and Slater was home to the first fully mechanized cotton-spinning mill in the United States. Slater's water-powered textile mill, funded by Rhode Island merchants William Almy and Smith and Obadiah Brown, began production in 1790 and three years later was profitable enough to justify expansion.[83] Slater's industrial revolution was not the standard experience of most early manufacturing attempts that Wansey observed, but perhaps Wansey would have reconsidered his depiction of US manufacturing had he ventured to Rhode Island.[84]

The business enterprise of Almy, Brown, and Slater called into being the Blackstone River Valley as the first manufacturing zone in the United States. It was home to small-scale spinners and weavers who, before the 1790s, mostly tinkered with machines, rather than making major changes in productive capacities. That changed with the immigration of Samuel Slater, and soon the partnership with Almy and Brown spawned similar textile ventures. While these mills would later be eclipsed by the factory system of eastern Massachusetts, their original business model of private-wealth financing and limited production—mostly yarn—proved modestly successful.[85]

Quaker merchant Moses Brown became interested in manufacturing and formed a partnership with William Almy and Smith Brown. Realizing that textile manufacturing would not be profitable unless they mechanized production, they acquired machines modeled after Richard Arkwright's famous water frame, but they lacked the ability to assemble the components in their proper static and dynamic interrelationships. Enter Samuel Slater, who was working as a factory manager in Derbyshire, England, when he was lured to the United States by the promise of financial remuneration for technical improvements. Once the United States ratified the Constitution and political stability seemed secure, employment in the new nation offered a viable alternative to manufacturing opportunities in Britain. Even as British statesmen and manufacturers disdained US manufacturing attempts, they jealously guarded their technology and their workers.[86] Attempts to quarantine their manufacturing processes, however, were no match for Americans' eagerness to import human and technological capital from Britain. Immigration offered the best solution for transferring technology, especially as US citizens lacked the technical know-how to operate machines that had been developed under a foreign system of measuring and counting.[87]

The United States' open immigration policies made possible Samuel Slater's role in launching a manufacturing system that allowed the young nation to compete with its former colonizer.[88] Slater arrived in New York in 1789, where he attracted the attention of Moses Brown. Slater told Brown he could successfully install spinning machinery for his firm, and they offered him all of the company's profits for a six-month trial period. The men formed a business arrangement by which Almy and Brown provided capital and Slater oversaw production.[89] Slater introduced and improved mechanized spinning in the Blackstone Valley, where workers still used hand technologies. The version of Arkwright's water frame he adopted enabled the firm to use the currents of the Blackstone to make yarn more quickly and more cheaply.

The technology Slater introduced gave it an edge over companies like the Worcester Cotton Manufactory that still used pre-Arkwright spinning machines. Variations of Arkwright's machine, which employed water power to turn cotton into yarn, had been constructed without success throughout the Northeast. Once installed, the machine increased the amount of yarn and thread the firm's twenty-three weavers transformed into cloth. Almy and Brown also constructed a dye house, which further distinguished the firm from earlier forms of textile manufacturing because it integrated additional steps of the production process within one firm. It required the importation of foreign dyestuffs, such as fustic (a yellow dye made from *Maclura tinctoria*, or dyer's mulberry) and logwood from South America.[90] Yet for all its appearances of modernity, it was still a relatively small partnership that relied on putting out to household producers and rarely operated at full capacity.[91] Slater's imported technology made possible the success of the first textile spinning factory in the country, but it alone was not enough to create industrial growth on a regional scale. Slater's factory model was not the model that would symbolize American industrialization, for within thirty years of its founding, it would be eclipsed by the Waltham-Lowell System.

Almy and Brown's marginal success suggests that American cotton manufacturing would yet require the stimulus provided by wartime embargoes on British goods. Demand for Almy and Brown's products was mixed. Consumers still preferred imported cotton cloth, but they were beginning to purchase more domestic yarn. Shopkeepers in Baltimore and Salem issued glowing reports of good sales and a desire to sell more; others offered the opposite reaction.[92] In order to sell cloth, the firm had to undersell household producers and meet the prices of British imports, a practice which sharply limited profits.[93] A. and A. Smith of Wickford, Rhode Island, reported that they could no longer sell the firm's cotton goods because the profit was so small, while another merchant found the "selling of yarns attended more trouble with greater interruption than [he] expected."[94] When merchants declined to offload Almy and Brown's products, the firm simply found other sellers. Their yarn usually found ready customers among women who engaged in household weaving, either for their own families or for sale outside the home, but this demand went only so far.[95] One merchant in Poughkeepsie, New York, informed Almy and Brown that he had been approached by a factory manager in Connecticut to dispose of his yarn, but he said he would continue selling for the Rhode Island firm because he preferred to do business with manufacturers who would allow him to sell at the lowest prices and still provide commission.[96]

Their customers were savvy; they expected that when the price of cotton fell, the price of yarn should fall, too, regardless of a firm's other economic considerations.[97] Almy and Brown could sell low because they, unlike others, had sufficient wealth to manage their business. Also, the firm was a small partnership, unbounded by profit-maximizing considerations for shareholder payouts. For these reasons, it survived into the nineteenth century, while firms like those at Worcester and Beverly failed.

Before the Embargo of 1807, there were only fifteen textile mills in the country, half of which were located along the Blackstone River and owned by Almy, Brown, and Slater; they primarily produced cotton yarn.[98] Of the sixteen mills in operation in Rhode Island at the time of the manufacturing census of 1810, two-thirds had started since the United States enacted prohibitive importation policies. One-fifth of all spindles in the United States were operated by one of Slater's factories.[99] During the 1807 trade embargo against Britain, G. K. Pitman, agent of the Providence Manufacturing Company, recognized that if normal trade with Great Britain resumed, British merchants would flood the market with cheaper factory goods, making it difficult once again for American manufacturers to compete.[100] During the 1780s and 1790s, profitable textile production was not yet feasible for the very reason Pitman laid out. The domestic market could absorb only so much yarn, and Almy, Brown, and Slater had successfully captured this market. Before the embargo, wealthy capitalists like Thorndike, who had lofty revenue expectations, preferred international trade, while aspiring upstarts could not make a living from a cotton mill.

And so, for the better part of two decades, Almy and Brown, along with a handful of others, were the sum total of the American textile industry. While Almy and Brown still operated at full capacity thirty years after its founding, the firm would not be the face of industry for long.[101] The War of 1812 would demonstrate to policy makers that clothing and blankets were military necessities.[102] Textile manufacturing was an economic imperative and it would, in time, receive more effective federal support. When it did, new competitors, namely along the Merrimack Valley, would expand and surpass the Blackstone Valley's Rhode Island System.

Before geopolitical realities created manufacturing opportunities and forced a more coherent industrial policy on the federal government, the burden was on manufacturers to promote their own importance. Beginning with the Pennsylvania Society for the Encouragement of Manufacturers and the Useful Arts, merchants and mechanics organized societies in cities from Boston to Baltimore to promote their usefulness to society. They were part voluntary associations and part industrial corporations that brought together wealthy merchants, skilled workers, and interested politicians.[103] While these societies produced more verbiage and fewer goods, they did much to advertise the values of manufacturing. The Providence Association of Mechanics and Manufacturers, for example, endorsed the ratification of the Constitution, recognizing that a strong federal government was necessary for trade protectionism, and pledged to supply Congress with information on manufacturing. Its members also urged manufacturers in other cities to found similar societies.[104] The most well-known of these manufacturing societies was the Society for Establishing Useful Manufactures in Paterson, New Jersey. When Wansey visited Newark, New Jersey, he dismissively mentioned the undertaking as another example of why America should stick to agriculture rather than industry.[105] It was the brainchild of Tench Coxe, who had assisted Hamilton in writing the *Report on Manufactures*.[106] Together with Hamilton, Coxe enlisted wealthy New York investors to establish an industrial center with mills for spinning cotton. Federal bonds helped fund the company, which received a corporate charter from New Jersey, along with preferential tax treatment. Nonetheless, it failed. Despite backlash against this public-private partnership, public support for manufacturing societies persisted. The Providence Association, for example, would continue to influence legislation through the Civil War.

These societies had gotten the message out: manufacturing was important for the new nation. In the early republican United States, manufactures of all kinds enjoyed political support. A producerist ethos that had developed during the Revolution meant that it mattered politically whether Americans or foreigners clothed the new nation. This moral construction of domestic manufacturing persisted in varying forms through the Civil War, when seamstresses and tailors would protest the contract-purchasing of military uniforms from "non-producers."[107] In the nation's first decades, though, the issue was not yet military uniforms, but general sustainability. Even those who favored an agrarian economy recognized the virtues of productive capabilities. James Madison, for example, in his plan for negotiating with Britain, advocated the exclusion of British manufacturers in favor of the development of cottage industry. And even opponents of Hamilton's plans for manufacturing and militarization understood that the federal government could help promote economic development. Secretary of Treasury Albert Gallatin, for example, believed that a national bank would enable the federal government to appropriate money expressly geared toward development projects and that the United States should lend money to merchants to help them compete effectively against the British in foreign markets.[108] Federalists and Jeffersonian Republicans alike recognized that both national security and aspiring manufacturers—their political constituents—depended on federal support. Although the nation's founders had different opinions on how economic equilibrium could be achieved, they all recognized that political independence, military security, and economic development were intricately linked.[109] Yet, for all the political rhetoric championing the small producer, the federal government would channel most of its resources toward the arms industry and offer little financial support to textiles until the tariff of 1816.

Guns for Soldiers

After leaving Worcester, Wansey headed to western Massachusetts, where he visited Springfield, too soon to see much more than a paper mill and some printing presses. He remarked on the "handsome" arsenal and powder magazine, but wasted no time speculating about arms manufacture. He admired the Connecticut River, "a charming river, winding, like the Thames, through a fruit valley."[110] Unbeknownst to Wansey, this river would power the American arms industry.

President Washington's second annual address warned Congress that they should not overlook the possibility of war, which could happen at any moment with nations "most concerned in active commerce with this country."[111] Calls for manufacturing centered on economic security, but there was a more tangible aspect to security that went beyond the physicality of American-made cloths and trinkets. Protection against threats from outside, and from within, was of paramount importance as the United States established itself as an independent nation.[112] The solution was guns. The armed citizen would give the early national United States a decided advantage over despotic European nations that did not trust their own civilians with guns. Guns were in many ways a moral imperative for citizens of the early republic. Republican thought dictated that society could be maintained only by citizens who were willing to protect it.[113] Although Americans generally believed that state militias offered the most virtuous form of protection for republican citizens, the disastrous defeat by the Western Confederacy of American Indians in 1791 proved the militia's insufficiency.[114] In 1792, the regular army was revived as the Legion of the United States, "having all the component parts of the largest army of any possible description, [and] prepared to meet every species of war that may present itself."[115] Regardless of their opinions of a standing army, American officials believed in the relationship between firearms and national security. Politician and diplomat Joel Barlow argued that the American experiment depended on the fluidity between soldier and citizen, who should be "obliged" to bear arms. Thomas Jefferson believed the gun should be one's "constant companion." And George Washington recognized that the surest way to preserve peace was to be prepared—with arms—for war.[116]

Output from two federal armories, supplemented with contract arms, would supply the new nation with the weapons it needed to fend off insurgent Americans, Indian neighbors, and belligerent European nations. In the early 1790s, however, the nation's arsenals, half-filled with old European models in need of repair, did not meet American military needs. To many it seemed that the "manufacture of military articles would become inconsiderable during the existing peace."[117] Local militias were supposed to supply themselves, but cash-strapped state governments found it difficult to afford contracts with private manufacturers. It was too much to require individual militia members to contribute their own guns, which were often left at home during the Revolution for the protection of soldiers' families, and whose cost represented about two weeks' worth of wages for a laborer.[118] These shortcomings aside, it was not a foregone conclusion that the United States would produce or purchase weapons for the nation's militia. In 1794, Congress recognized that "the principal defects in the existing provisions for arming the militia consist in the want of a competent source of supplying the militia" and "resolved that further provision ought to be made by law for arming the militia of the United States."[119] Yet even in 1798, after the federal armory at Springfield had been established, congressmen still debated how and when militias should be armed: Speaker of the House Jonathan Dayton thought it would be "very inconvenient for the United States to become the retailers of arms [to the militia]" and Congressman George Thatcher of Massachusetts questioned why the "General Government should be put to the trouble and expense of providing arms for the state governments." Thomas Claiborne of Virginia countered that "at this time it was proper that arms should be put into the hands of the people and that those who are not able to pay for them should have them given to them."[120] It is not surprising that support for federal arms fell along regional lines, as many southern militias, Virginia's included, could only arm less than a quarter of their men.[121] Even as southerners often bemoaned federal intervention, they sought government assistance in arming white citizens, whose use of guns would quell slave insurrections.

Congressional squabbling reflected the local realities of state purses and voters' preferences, but federal officials advocated policies based on national security concerns. Secretary of War Henry Knox advised that "sound national policy" dictated that the nation produce its own arms, maintaining in its arsenals a minimum of one hundred thousand stands of arms (which included the musket and its bayonet and ramrod, both of which were fit to the weapon). Knox recognized that while manufacturing the arms might at first cost more than importing, the cost could not compare "with the solid advantages which would result from extending and perfecting the means upon which our safety may ultimately depend."[122] In 1794, the same year that Washington's administration sent the national militia to Pennsylvania to quell domestic unrest against excise taxes, Congress passed a bill to create and staff two federal armories, allotting $22,865 for construction expenses and $143,640 for the purchase of additional arms and ammunition.[123] Complete and consistent statistics for the early national period are hard to come by, but spending on the military began to constitute anywhere from 20 to 40 percent of total federal expenditures.[124]

An Act of Congress in 1792 gave the president power to select two sites for the nation's federal armories, and Washington chose Springfield, Massachusetts, and Harpers Ferry, Virginia. Springfield was an ideal site on which to construct a federal armory but a problematic one nonetheless. As Wansey had observed, it was already home to an arsenal and powder magazines, "both handsome new brick buildings."[125] It also boasted a dozen small workshops and storage buildings and a slew of barracks. It stood at the crossroads of several interstate transportation routes and had access to natural resources and raw materials from the mixed-wood forests of middle New England and a nearby iron foundry in Salisbury, Connecticut. Additionally, its location sixty miles north of the mouth of the Connecticut River made it safe from naval attack. The site, though, was on a steep-sided plateau with limited access to water power.[126] The Mill River, one of the larger streams that joined the Connecticut River, was chosen as the power source, but its limited drainage basin meant that it often experienced both floods and droughts. Also, although the federal government already owned the land on which the storage buildings were located, the government would have to purchase much more land in the area if it wanted to build the dams needed for generating sufficient power to produce arms on an industrial scale. The site's access to raw materials, iron and steel, skilled labor, and good roads, however, seemed to outweigh its natural limitations. With public funds, the federal government could purchase additional plots of land over the next several decades and establish what would become by the mid-nineteenth century the world's largest arms-making institution.[127]

On June 22, 1795, the federal government bought its first acre-and-a-half parcel of land for $400 and commenced production.[128] Workers, under the oversight of a superintendent and two master armorers, repaired old arms and manufactured the nation's first public musket, known as the Model 1795 musket.[129] This .69 caliber flintlock required the user to load powder from the muzzle end, followed by a lead ball rammed down with the ramrod. It was modeled after the French Charleville, which became the standard US musket until the War of 1812. The Charleville fired a smaller round than the British Brown Bess, but it had a longer range and was more accurate. Secretary of War Timothy Pickering deemed it important to establish some degree of standardization for the military's weapon use and, because this model was considered superior to any other European gun and had been used in great numbers during the Revolution, it seemed logical to make it the US standard.

The United States had imported most of its guns from France during the Revolution, but in the 1790s war with its old supplier seemed inevitable.[130] The Springfield Armory could not at first meet War Department needs. During its first few years of production, workers at the armory manufactured guns under an apprentice system in which old handicraft techniques were used. They produced only two hundred forty-five muskets during the armory's first year of operation and eight hundred twenty-five its second year, plateauing at a little over one thousand muskets for several years after.[131] In order to expand production, the federal government established four new factory sites on the Mill River, dividing the establishment between shops that used water power and those that used hand power.[132] These new establishments enabled the armory, under the direction of the second superintendent, Henry Morgan, to introduce labor-saving machinery and divide the workforce into four categories of labor: barrel makers and forge men; filers; stockers and assemblers; and grinders and polishers.[133] In order to keep pace with the higher wages paid by private contractors to their employees, the War Department also started paying workers more. Officials figured that higher wages would increase the number of quality workers at the armory and that over time wages would go down. After initial struggles to construct machinery and determine which men should be placed in which positions, efficiency at the armory began to increase. Production changes cut the number of days it took each worker to manufacture one gun from twenty-one to nine and reduced the public cost of a musket from $13.17 to $9.29. The 130- to 150-man workforce was soon able to complete one day's work in half a day.[134]

Even with improvements in production efficiencies, though, the armory could not at first produce enough guns to arm the military. It eventually produced thousands of muskets each year—ten thousand by 1810—but in the 1790s it manufactured fewer than one thousand annually.[135] The War Department turned to private contractors, whom it hoped could supplement the limited output of Springfield Armory. This change in policy, though, reflected wishful thinking, since arms making remained an underdeveloped industry.

In fact, it had completely escaped the notice of Wansey. Upon visiting Middletown, Connecticut, which within a decade would be a major center of private arms contracting, the only noteworthy thing he saw was his first maple sugar tree.[136] In the 1790s, gunsmithing in the United States was still a small-scale, specialized trade. In the *Boston Directory* in 1800, for example, there were only four gunsmiths listed and each produced a small number of arms per year. Gunsmiths spent about a month of labor on each weapon, often forging the barrel, assembling the gunstock, and completing the grinding and filing tasks themselves. Much of their work was dedicated to repairing old firearms.[137] Labor was expensive, and consumer demand did not warrant a heavy capital investment in the enlargement of gun factories. Most families purchased only one gun for their households, if that. General stores and wholesalers' inventories illustrate this: they were filled with foodstuffs, candles, and clothing items, not muskets and rifles.[138] The private merchant ships and occasional privateering and filibustering expeditions that needed weapons on board did not provide reliable demand, either.

Officials in the War Department believed that national security merited the expense, and even a bickering Congress understood security issues.[139] The Indian campaigns that carried over from the Revolution wound down with the signing of Treaty of Greenville in 1795, but the nation remained on high alert. Policy makers did not want another embarrassing defeat like the one General Arthur St. Clair experienced in the Northwest Territory in 1791, when militia deserted in droves and American soldiers were given shoddy coats and hats and shoes that "lasted not more than four days."[140] Indeed, Secretary of War Timothy Pickering argued that despite recent victories, the United States must not countenance any more ill-prepared conflicts with Native Americans on the frontier, like those a few years earlier.[141] The terrors of border conflict for white Americans were very real, even for those who lived closer to the seacoast than the frontier. Depictions of Native American violence were ubiquitous, and many prominent Americans read accounts that warned, for example, that "our frontier settlements must ever lie at the mercy of the savages."[142]

In addition to never-ending border warfare, the United States routinely dealt with commercial and naval conflicts with European nations. In the late 1790s, the United States prepared for war with France, with Congress allocating $800,000 for munitions after negotiations over naval aggression failed.[143] Even after the Quasi War against France ended with an agreement that reduced French incursions against American shipping, a US congressional committee on "the system of national defense" trumpeted the importance of security. Despite its recommendations for some reductions in military expenditures, it advocated that "no such material change in the state of the foreign relations of the United States has happened, as would justify a relinquishment of any means of defense heretofore adopted by Congress, but that the national honor and interest, in the present posture of affairs, make it prudent and necessary to continue to be prepared for the worst event."[144]

The same year that Wansey visited the United States, the federal government issued contracts for seven thousand muskets, all of which were made in Pennsylvania. Four years later, it issued a larger round of contracts for forty thousand stands of arms at $13.40 each.[145] Federal officials believed these could be made more cheaply at the public armory, but the armory did not yet have the capacity to turn out the required quantities.[146]

All of the muskets were to be modeled, like those at the federal armory, after the French Charleville musket.[147] No longer purchasing weapons from France, the US government instructed its contractors to make weapons worthy of its foe's military. Thomas Jefferson admired the superiority of French arms making, and the War Department ordered two volumes of a French guide to manufacturing weapons, complete with tables on standardized measurements.[148] While serving as minister to France in the 1780s, Jefferson visited the workshop of Honoré Blanc, an inventor and military gunsmith who had developed a new method for manufacturing the musket's flintlock (the firing mechanism used on rifles and muskets from the seventeenth through the nineteenth centuries). Blanc was able to assemble the flintlock from an assorted array of interchangeable parts.[149] Jefferson, deeply impressed by the manufacturing processes he witnessed—which would later in the nineteenth century become known as "interchangeability"—encouraged Blanc to emigrate. When that did not happen, Jefferson settled for shipping a selection of French muskets to the United States for study.[150]

Eli Whitney, of cotton gin fame, is often considered the first American to bring French interchangeable production methods to fruition. In the 1790s, however, nothing about Whitney suggested this would be the case. When Wansey toured New Haven, admiring the exotic snake skins at Yale College's library, Whitney was embroiled in patent disputes over the cotton gin he had allegedly invented.[151] There was certainly no sign that New Haven, "a very neat pleasant town," whose only notable manufacturing included a cider mill and a struggling cotton and woolen manufactory, would become a center for arms manufacture.[152] Several years after Jefferson returned from France, Whitney was struggling to establish an arms manufactory. He had no experience making guns, but he thought government work would prove more profitable than his previous ventures. He was friendly with Secretary of Treasury Oliver Wolcott, to whom Jefferson had described Blanc's manufactory and sent a pamphlet on French arms-making. Wolcott forwarded this pamphlet to Whitney, suggesting that the information would "help establish the manufacture on a permanent footing."[153] Whitney made the most of the privileged information he received about French interchangeability, even as he snidely remarked to Wolcott that the French pamphlet "appeared calculated to mislead as much as to instruct."[154] In 1798, Whitney traveled to Washington, where he reassembled the scrambled parts of several muskets in front of an influential audience that included President John Adams and Vice President Thomas Jefferson.[155] Interchangeability trickled slowly into American industrial consciousness, and the uniformity of weapon parts did not become a federal priority until after the War of 1812, but Whitney, as purportedly the only private manufacturer familiar with cutting-edge French techniques, secured the largest of the 1798 government contracts. His contract for ten thousand guns vastly exceeded the other contracts, most of which were for fewer than one thousand stands of arms.[156]

Not only did Whitney receive the largest order; he was the only contractor to get a cash advance. Even Daniel Gilbert, of Mansfield, Massachusetts, who contracted for two thousand muskets, complete with bayonets and ramrods, was not paid until after he delivered his order.[157] In addition to the cash advance, the War Department promised to pay for several storehouses on Whitney's property.[158] Whitney had convinced War and Treasury Department officials that he could successfully transition from cotton gins to guns—before he had even constructed an armory. Decius Wadsworth, the War Department's superintendent of small arms, wrote that Whitney could manufacture arms "superior to any musket for common use yet fabricated in this country" and was "highly deserving of national patronage."[159] The patronage Whitney received, though, was more the result of his personal and political connections.[160] He was able to recruit reputable Connecticut men to serve as sureties on his government bond, and Wolcott promised Whitney he would "guard against too much competition" from other arms manufacturers' proposals.[161]

Reputation and federal aid still did not guarantee that Whitney would fulfill his quota on time. In theory, contracts between producers and the federal government would be mutually beneficial relationships. According to Whitney, "it is in the interest of national security to train men to make arms . . . in preference to worthless vagabond foreigners."[162] Arms production could serve patriotic aims while also lining pockets. Yet Whitney's experience demonstrates a major reason the first contractors were unable to meet the terms of their contracts: most were enticed by federal patronage but unprepared for industrial-level production. Whitney asserted that he could manufacture muskets more cheaply than they could be imported by employing machinery for some of the process.[163] Although he had not yet built a factory of a size that would enable him to undertake the manufacture of ten thousand muskets, he "had assumed that the [previous] secretary of treasury understood this." Although construction occupied Whitney's time, he assured the secretary of treasury that he was working hard, despite delays in the delivery of muskets, and that he would "effect the contract with some profit to myself and to the satisfaction of the government."[164]

Whitney's writings reflect a preoccupation with extracting the greatest possible output at the lowest cost from a limited number of unskilled and semiskilled workers in order to complete his contract.[165] He cautioned federal officials that the establishment of another factory near his own would raise the price of labor, which would be detrimental to profits and would make it difficult to "keep good order among the workmen."[166] Whitney's obsession with cost efficiency was one of the downsides of relying on private contractors. In many ways, Whitney was more interested in milking the contract system than in reaching production quotas.

After much struggling, Whitney was allowed to take until 1809 to fulfill his obligations, but the rest of the fifteen contracts from 1798 (contracts with other manufacturers) were terminated after several years, some of them incomplete. If Whitney could not fill his contract with all the help he received, it is no surprise that others could not either.[167] Part of the reason for this shortage was no doubt because of kinks in the system. The War Department, for example, experienced difficulty getting patterns of the Charleville musket to all their contractors, and in 1798 none of the six Vermont arms makers had received a pattern.[168]

The reason federal officials ultimately blamed for the shortage was the method by which the government found suppliers. While the government eventually contracted only with familiar and reputable men, the Treasury Department initially selected would-be manufacturers from those who answered an advertisement. Many made promises they could not keep.[169] Less than a year into the contracts, the Treasury Department expected noncompliance. One manufacturer sent in his proposal after the requisite number of muskets had already been contracted for (contracts were essentially first come, first served), but, recognizing that contract failure was likely, Secretary of Treasury Wolcott entered into an agreement with the man for five hundred muskets to make up for the anticipated supply shortages.[170] During the War of 1812, the government would forgo this practice of soliciting arms from private parties who had not been vetted by federal officials, and instead would dole out cash advances to only a small cadre of reputable contractors in order to ensure adequate supply.

But the main reason the contracting system did not work was that industrial arms production was still in its infancy. While the system of soliciting contract bids from the civilian market worked for the Navy Department with the shipbuilding industry, arms making was not yet the developed industry that shipbuilding was.[171] Manufacturers needed enough cash to induce them to produce large quantities of government-standard arms, rather than the garden-variety agricultural implements that provided the livelihood of many gunsmiths. Simeon North and Nathan Starr, for example, returned to manufacturing farm tools at their Connecticut factories as soon as their 1798 contracts ended. Their choice suggests that without government consumption, manufacturers had little incentive to produce guns.

Federal officials, however, persevered. Even as military conflict with France seemed less a threat, the War Department cautioned Congress that "it can never be wise to vary our measures of security, with the continually varying aspect of European affairs."[172] The nation must always be prepared for war. As the United States entered the nineteenth century, the War Department cultivated relationships with private contractors to augment the federal armories and directed newly made-in-America guns to an expanded military, while the Treasury Department dispensed funds to the two national armories and executed contracts with private manufacturers.[173] In the hands of the executive, the arms industry continued to grow. The legislature had given to the War Department the right to employ as many people as it saw fit in the production of public arms and to oversee the manufacturing and provisioning processes.[174] Any misgivings about the military became inconsequential as the nation, led by the War Department, prepared for war at any time.

Rewards of "Independency"

Americans eventually proved both Lord Sheffield and Henry Wansey wrong. When Tench Coxe issued "A Statement of the Arts and Manufactures" in 1812 for the Treasury Department, he celebrated New England textile manufacturers for rendering "green seed cotton"—its fiber recently liberated by the cotton gin and an increasing number of slaves—into a productive material domestically.[175] In many ways, Alexander Hamilton's vision two decades earlier had come to fruition; cotton was no longer merely an export, but a raw material that could be spun and woven into cloth that offered to free Americans from economic dependence on England. By 1812, too, the United States no longer needed to import its weapons from France. Coxe reported that "the expense and trouble of a judicious and rigorous inspection . . . have made favorable changes in the condition of this important branch of our manufactures."[176] From a federal standpoint, the money and time spent on the production of guns was worth it. The War Department had entered into its second major round of contracts and was prepared to arm its soldiers for war against England. The United States had come a long way since it had assessed the state of its manufactures and embarked on a national plan to produce its own weapons twenty years earlier.

The industrial mission, though, was far from complete. While public funds had created an arms industry, military supply inadequacies persisted on the eve of war with Britain, and the nation still required European imports to clothe its soldiers. Up through the first decade of the nineteenth century, the majority of merchant wealth was still tied up in international commerce. Men of Israel Thorndike's status continued to find Cuddalore ginghams more profitable than Rhode Island bedtickings.[177] But once the federal government made textile production a greater concern, this would change. Textiles, like arms, were a matter of national security, and US statesmen would start to do more than parade around in suits made of domestic cloth. Thirty years after independence, manufacturing was no longer a question. Instead, American policy makers asked how best they could channel federal resources into public and private industries to ensure national prosperity and unchallenged dominance over the continent. Wansey and Sheffield may have misjudged Americans' ability to manufacture, but French revolutionary Jacques-Pierre Brissot had it right when he said that "[American] industry is sure to receive the reward of independency."[178]

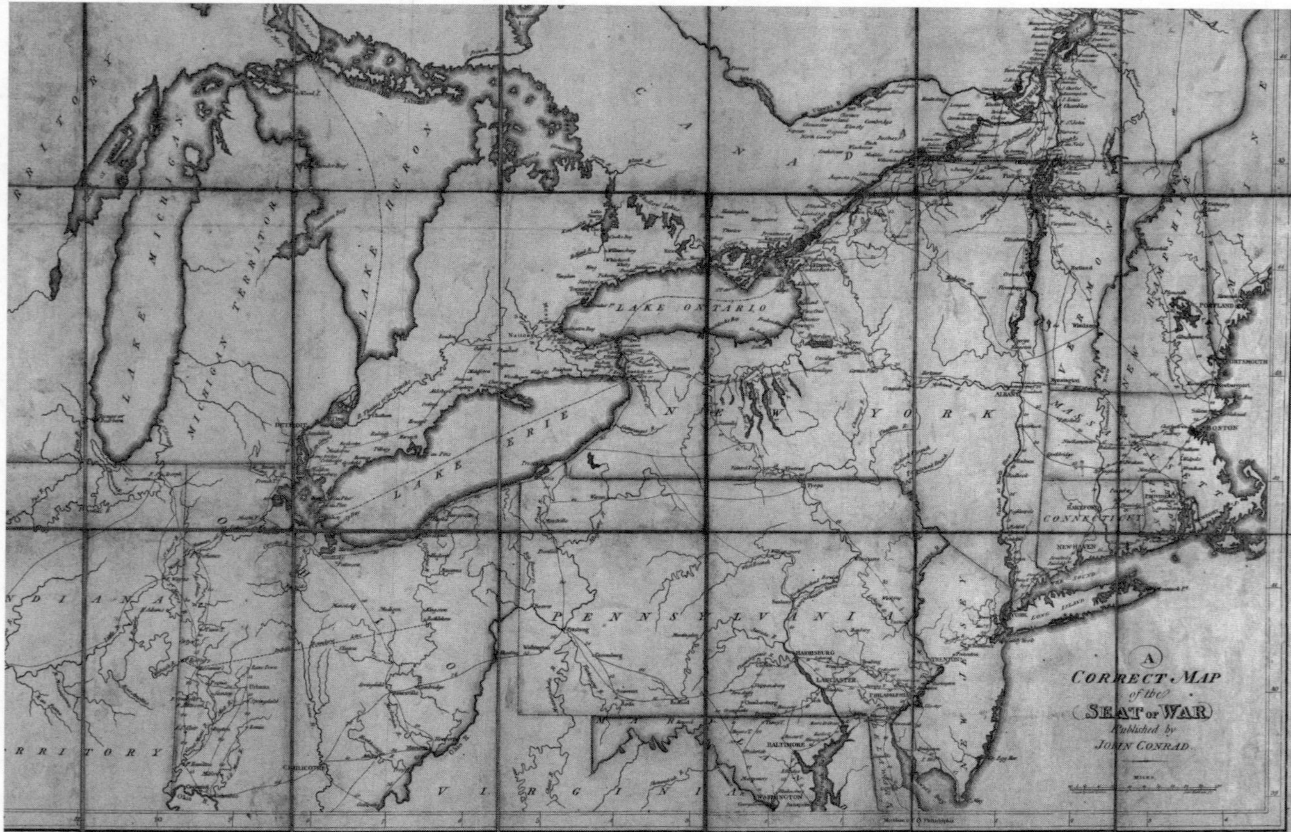

The War of 1812. From Samuel Harrison, *A correct map of the seat of war/drawn by Samuel Lewis* (Philadelphia: John Conrad, 1812).

Courtesy, The Library Company of Philadelphia

[3]

Embargo and War

In 1810, J. and S. Hindsdill built a woolen mill in Bennington, Vermont. The southwestern Vermont town was home to paper, iron, furniture, and textile manufacturers who benefited from the generous water power offered by the Walloomsac River and Paran Creek, as well as from proximity to rich natural resources. Natural advantages aside, the Hindsdills, like many aspiring manufacturers, found the economic promise of nonimportation a more compelling reason to go into business. Denied British imports, American consumers would likely turn to local manufacturers in towns like Bennington for their purchases. Despite a profitable first few years, however, the Hindsdills "abandoned the business of manufacturing" ten years after the company's founding, citing insufficient government encouragement.[1] This trajectory of the life and death of one mill's operations was typical of many other small textile factories over the course of conflict and resolution with Britain. Yet it is not the only manufacturing story to be told in the context of the War of 1812. While the Napoleonic Wars gave rise to a number of small textile factories throughout New England that did not survive peace, they also initiated the nation's first large-scale textile industry and a thriving arms industry that did. The embargo and war made evident that it mattered whether a factory's capital came from mercantile wealth, the federal government, or locally pooled resources.

On June 18, 1812, President James Madison signed a declaration of war on Great Britain, further angering a New England mercantile community that already opposed US commercial policies against Britain. What seemed disastrous for trade offered an opening for domestic manufactures. Without competition from British manufactured goods, American goods could dominate the market for cloths, firearms, and other wares. Additionally, military demands for war materiel created opportunities for supplying the army. Whether the War of 1812 is characterized as a "second war of independence" or as a civil war among various American groups, the United States' war-making bolstered industrial development.[2] When we understand the War of 1812 as part of a European mercantilist competition for "geopolitical, commercial and economic hegemony," as scholars have conceptualized Britain's participation in the Napoleonic Wars, we see how protection against imports and favorable fiscal treatment for strategically important manufactures sparked long-term industrial development, even as it led to postbellum economic contraction.[3]

Scholars studying such conflicts as the Napoleonic Wars in England and the Civil War in the United States have looked for patterns in the relationship of government policy to the production of specific goods. For example, Mark R. Wilson's study of the Civil War tent industry posits that many military supply industries, especially textiles, were characterized by extensive networks of workers and suppliers, in contrast to the centralized, capital-intensive arms industry. Decentralized production arrangements, in which far-flung factories and workshops sold to merchant middlemen, dominated the procurement of most wartime supplies, save for arms and ships.[4] Wilson's notion of the military supply economy during the Civil War can be applied to the early national context as well. While federal agents relied on a combination of merchant suppliers and widely spread small textile firms to clothe the army, they depended on an increasingly concentrated core of arms manufacturers. The arms industry needed public capital in a way that the textile industry did not. Textile manufacturers did, however, depend on the wartime economic stimulus—a stimulus that produced uneven effects immediately following the war. Many companies that started in the context of the embargo and war suffered during the postwar resumption of trade. Likewise, not all contractors received peacetime renewals. It was these differences that shaped the course of the Industrial Revolution in America, an epoch of economic development that cannot be understood without giving government war-making powers a central role.

Preparing for War

The first act of retaliation against Britain for offensive naval actions was the Non-importation Act of April 1806. Although ineffective, it marked the first step toward excluding British goods from American markets. Subsequent commercial legislation did little to inflict material pain on Britain, especially as smuggling proliferated and customs officials had no real incentive to enforce the law. But what historian Gautham Rao describes as "perhaps the most radical piece of market regulation in United States history" provided space for industrial growth.[5] Even as the federal state struggled to enforce the embargo, British imports did decrease, and merchants, suffering from diminished trade, looked to factory investment. Additionally, in the context of possible warfare against Britain, the federal government took measures to ensure that it could supply itself. In 1808, Congress passed an Act for Arming the Militia, providing $200,000 per year to provide guns for the state militias, which, under the Militia Act of 1792, had been required to supply themselves. And in 1810 it commissioned a national report on arts and manufactures and began soliciting manufacturers for improved goods.[6] This federal concern with preparedness resulted in haphazard supply, but as the nation prepared for war its industry expanded beyond what had seemed possible at the turn of the century. The arms industry, the beneficiary of annual appropriations and government contracts, generated sufficient output for military needs. The textile industry, on the other hand, experienced countless new mill openings, but did not yet produce the quantity or quality of cloth necessary to clothe the army. Limitations aside, commercial warfare made possible the growth of industry, even as it hindered general commerce. It was this period of retaliation against Britain that established manufacturers' dependence on the government.

Preparing for War: Arms

The founding of two federal armories in the 1790s had not fixed the nation's firearms supply problems. It took well over a decade to establish adequate infrastructure and management, and there were often supply shortages as War Department officials scrambled to procure materials for weapon making.[7] The Springfield Armory manufactured less than five thousand muskets per year before 1807, and not more than ten thousand until 1811.[8] Harpers Ferry performed even worse, and in an 1806 survey of "firearms fit for use," it possessed only a quarter as many muskets as Springfield.[9] Additionally, there was no comprehensive plan for providing arms to the nation's soldiers. Prior to the War of 1812, some states, such as New Jersey, Pennsylvania, and Virginia, issued state-owned muskets to active militia; others required all members to purchase their own arms.[10] Many of the state-owned arms were European models left over from before and during the Revolution, and southern states especially struggled to arm even half their militias.[11] Some states turned to private contractors to bolster their supply. New York, for example, contracted with Eli Whitney to help supply the roughly 40 percent of militiamen without guns.[12] For a short time, Whitney was on both state and federal payrolls. This patchwork system did little to help cash-strapped states or to engender militia standardization. All told, many of the 113,501 muskets the nation owned in 1806 were decades old and in need of repair.[13]

These inadequacies, compounded by the impending hostility against Britain, prompted Congress to debate how best to ensure the existence of a well-armed militia in the event of war. Although some congressmen wanted each militiaman to supply himself, believing federal aid would encroach on state military prerogatives, the majority agreed that the federal government needed to "create a system of supplying the nation."[14] As one Virginia congressman wrote to his constituents, the army "may soon require a hundred thousand men instead of four thousand."[15] The 1808 Militia Act reflected a compromise between those who wanted the federal government to supply the militia and those who wished to allow states to contract with private armories.[16] The act authorized the president to sell federally owned weapons to individual states that wished to purchase them, while also appropriating $200,000 to provision the various state militias with arms, usually between fourteen thousand and fifteen thousand muskets, or the equivalent (muskets were the predominant firearms, but this number included a combination of rifles, pistols, and swords), per year.[17] As part of the compromise, these arms would come from private contractors. Although the federal government purchased military supplies in the "free" market, it was wary of purchasing arms that did not conform to government standards.[18] When the government contracted with a private firm, it could supply that manufacturer with a pattern arm to replicate, while safeguarding itself against financial loss by requiring contractors to enter surety bonds.[19] Beginning in 1808 the purveyor of public supplies made five-year contracts with small arms makers and offered, for example, $10,075 up front for the promise to supply ten thousand stands of arms to the United States.[20] Before this time, no private arms maker operated a large-scale factory, something that required ample investment capital and reliable demand. Private investors, though, shunned arms manufacturing, and the average male citizen only needed one gun. That was why even manufacturers who had received federal contracts in the 1790s halted arms making until the next round of contracts in 1808.[21] By then the government realized cash advances were the only way to induce manufacturers to maintain an expensive and risky undertaking. As purveyor of public supplies, Tench Coxe had recognized in 1807 only advance-sum contracts could "excite and promote the small arms manufacturing and bring the business to settled form."[22]

Where and how it settled, though, was not yet obvious in 1808. The first twenty-two arms contractors were scattered throughout the Mid-Atlantic and the Northeast, but by the early 1810s, this number had dwindled to a small cadre of arms makers, the majority of whom were located in western New England.[23] It was with these contracts that the federal government eventually cemented its relationship with a select group of contractors and developed the Connecticut Valley as a center of federal arms production. The reasons the government committed to manufacturers in this region in favor of those elsewhere include proximity to the Springfield Armory, regional expertise, and perhaps the ability to influence novices.

Throughout the eighteenth and early nineteenth centuries, the most established gunsmiths were in Pennsylvania. They specialized in rifles for frontiersmen and custom work for the civilian market. Rifle making was relatively unknown in New England because civilian purchasers there did not have the same frontier needs as those in western Pennsylvania or Kentucky. Muskets were the norm there, and muskets were still the preferred weapon for military use.[24]

In general, however, New England did not have the long-established clusters of gun craftsmanship—musket or rifle—that Pennsylvania had. This relative dearth of expertise was not necessarily a problem for the federal government and actually may have been seen as an advantage. An experienced gunsmith came to contract agreements with his own sets of expectations and requirements. William Henry, a prominent Pennsylvania gunsmith, for example, traveled to the office of the secretary of war on January 2, 1812, and "particularly pressed the subject" of his and other area manufacturers' desire to deviate from the pattern muskets supplied by the federal government. Henry left the meeting with the impression that Secretary William Eustis approved their deviation and would compensate them for whatever additional expense they incurred improving the pattern. A year later, however, Henry and other Philadelphia-area merchants petitioned the secretary's office for payment they claimed they were due. This sort of issue was a nuisance to federal officials engaged in wartime operations. How could they successfully supply troops if they had to rely on manufactures who "did not consider themselves bound to the pattern," and "entirely free to act on their own judgment?"[25]

Experience was less important than malleability and potential. Western New England did not have the celebrated arms manufactories of Pennsylvania, but it boasted expert craftsmanship. The region was home to a large number of skilled artisans who made fine combs, spectacles, and clocks.[26] Also, as we have seen, these craftsmen included toolmakers who became regular arms manufacturers with government assistance. Comb- and clockmakers enjoyed a civilian market, but mass-produced weapons required robust military consumption. The best way to capitalize on these loans was to cultivate a cadre of manufacturers in one river valley, where the federal government could most effectively exploit both geographic and human resources.

Historian Theodore Steinberg has shown the importance of water in driving economic and industrial change, as capitalists sought to control water power and colonize nature.[27] In the same vein, the federal government exploited the Mill and Connecticut Rivers for water power and transportation: its employees built dams to generate energy for the armory's machines, and they shipped lumber, iron, and cases of guns up and down the river from Vermont to Long Island Sound. When water power around the Springfield Armory proved inadequate owing to flood, drought, or delays in dam construction, the War Department relied on their contractors' access to water power. Geographers and economists have analyzed the confluence of human capital, investments, government policies, and labor mobilization, among other things, in creating "new industrial districts" like today's Silicon Valley and Route 128 outside Boston. Geographer David R. Meyer has applied this framework to "technology districts" in New England in the early nineteenth century, arguing that the textile and arms industries developed from the proper combination of human and financial capital, which then attracted further investment.[28] The development of the federal armory in Springfield and the armories of the surrounding area was mutually reinforcing. The federal government had chosen this region not only for its strategic location on a plateau above the Connecticut River and its proximity to several major iron manufactories, but also for its technological expertise, especially in machinery and metalworking.[29] Additionally, there was a relatively prosperous population of farmers and businessmen who provided indirect capital to arms makers by endorsing notes for loans and posting bonds for federal contracts.[30]

The War Department would come to depend, for example, on North's and Starr's locations in Middletown, Connecticut. This historic port, situated on the western bank of the Connecticut River, just before it narrows and turns sharply east, offered water power and shipping advantages at only the cost of payments to its contractors there. This was especially helpful as the government improved facilities at Springfield by constructing dams and fire-retardant buildings. As federal officials tapped into the human capital in the Connecticut Valley, the region's manufacturers in turn benefited from proximity to government production. Unlike the small gunsmiths located in urban centers, the arms makers in this area received renewed contracts by virtue of their position in a technology district. They also received more timely remuneration during the war and did not have to petition the government for payment, as those in eastern Massachusetts did.[31] It made most sense to work with contractors located within easy travel of one another, especially because the government usually hired only one or two inspectors.[32]

By 1810, western Massachusetts produced more small arms than anywhere else in the Northeast.[33] In his response to the 1820 Census of Manufactures, gunsmith Lemuel Pomeroy of Pittsfield, Massachusetts, celebrated the government's decision "to manufacture their own weapons of defense within the United States by which means a good and perhaps sufficient number of the first rate gunsmiths are to be found to fabricate arms of the first quality as fast as the government may deem it necessary to procure them." The decision made it possible for him to keep his factory, which opened in 1809, in continuous operation.[34]

The War Department's utilization of the natural and social relationships of the Connecticut River Valley created a technology district that would serve national security for over a century. When the War Department entered into new contracts to fight Great Britain, the greatest number of contracts went to three Connecticut manufacturers: Eli Whitney, Nathan Starr, and Simeon North. Eli Whitney received a contract for fifteen thousand muskets in 1812, Simeon North for twenty thousand pistols in 1813, and Nathan Starr for ten thousand swords in 1813.[35] Of the three, Whitney had the strongest reputation as an arms manufacturer, but even he was a novice compared to men like the Henry family in Pennsylvania. As government cash flowed in and weapons flowed out to American troops throughout the eastern half of the continent, the Connecticut Valley made possible the United States' first major war against the Old World.

Preparing for War: Textiles

As the federal government cultivated gun production in western New England, it took a different approach with the textile industry. In the spring of 1806, Congress passed an "act to prohibit the importation of certain goods, wares and merchandise," which limited imports of woolen cloth and ready-made clothing from Great Britain.[36] While not consistently enforced until the end of 1807, restrictions on British imports offered American textiles a chance to succeed in markets usually dominated by British goods.[37] Even as the Non-importation Act and subsequent shipping embargo caused financial distress for many Americans in commercial centers, making it harder to sell goods of either foreign or domestic manufacture, textiles seemed a viable option for individuals seeking new business prospects.[38]

Furthermore, raw cotton was cheaper than it had been, owing to an increase in domestic production made possible by enslaved labor; as historian Edward Baptist has contended, industry was "built on the backs of enslaved peoples."[39] While the human torture that Baptist documents increased throughout the antebellum era, developments in cotton processing techniques, like the cotton gin, as well as federal policies that expanded the slave frontier, had drastically increased the supply of cotton in the United States. The Louisiana Purchase of 1803, in which the United States acquired vast amounts of territory for growing cotton, could be considered one of the earliest forms of government "support" for industry. Indeed, Baptist points to this purchase as the start of a so-called second slavery that would generate tremendous economic growth for the United States.[40] And as plantation owners began to experiment with more brutal forms of slave management, northern factory owners reaped the benefits of increased supply. They no longer needed cotton imports from Surinam, India, or Turkey.[41]

Hopeful upstarts enthusiastically took advantage of cheap raw materials, but commercial warfare against England provided the real impetus for industrialization. Britain was the largest consumer of American cotton, and its manufacturers churned out better, cheaper cloths than their American counterparts. The embargo provided the space for Americans to try their hand at what had largely been a British undertaking. In the context of federal commercial policy, new factories opened and established firms adapted to a market saturated with greater numbers of domestic producers.

Caroline Ware, whose 1931 study of early New England textiles remains one of the most thorough overviews of the industry, argued that price-fixing among manufacturers provided the chief stimulus for new textile factories. Her account rendered the embargo's effect negligible at best and detrimental at worst, a conclusion reached on the records of a single firm. Ware's proof centered on the fact that Rhode Island firm Almy and Brown had expanded prior to the Embargo of 1807. While Ware rightly pointed to the fact that the commercial restrictions destroyed the purchasing power of many of their customers in Boston and Portland, this was by no means ruinous for business. Almy and Brown simply sought new clientele in other seaboard cities like Philadelphia and Baltimore, as well as in the growing west.[42] In Ware's analysis, the industry expanded not because of new market opportunities, but because established manufacturers like Almy and Brown worked hard to keep prices stable despite market changes.[43] These manufacturers, though, had always engaged in some form of price-fixing, so it seems a stretch to credit the fact that beginning in 1806 new mills began to appear in Coventry and Cranston, Rhode Island, Paran Creek, Vermont, Stonington, Connecticut, and Middlesex and Worcester Counties, Massachusetts. These enterprises did well through the war, diminished consumer purchasing power aside. New selling houses devoted to vending New England yarn sprang up in major commercial centers. Yarn wholesalers in Philadelphia, for example, gave Almy and Brown a run for their money, underselling the Rhode Island firm in its Vermont markets. This competition forced Almy and Brown to adjust by offering a discount for payments in cash, by increasing the commission they paid, and by guaranteeing a specific assortment of goods.[44] It was not until foreign cloths reappeared in American markets following the Treaty of Ghent that manufacturers began reporting reduced sales, some even ceasing operations.[45]

The embargo also sparked military demand for domestic textiles. Reduced European imports meant that the War Department had to turn to domestic manufactures for some of its supplies. This would not have been the case if the government had listened to Secretary of War Henry Knox back in 1790 when he laid out the War Department's plan for organizing the federal militia. In his report to Congress, he calculated that clothing, the first and most expensive item on the supply list, would cost about $800,000 for three years' worth of uniforms for 325,000 men.[46] Great Britain and Russia received the bulk of that public expenditure by providing most of the US army's uniform-quality cloth. Instead of cultivating relationships with domestic textile manufacturers, the government employed supply contractors who often bought imported cloth from auction houses.[47] Commissioned tailors in the Philadelphia area then stitched together army uniforms with the imported materials. This system changed with the embargo and wartime commercial regulations, but army-grade cloth continued to show up in auction houses.[48] Through the 1810s, New York auction houses served as an important depot for British manufacturers and military contractors, who preferred quick, bulk sales at low prices to auction houses, rather than annual orders from traditional merchants.[49] Even after the first Non-importation Act in April, 1806, Purveyor of Public Supplies Tench Coxe continued to purchase army cloth from Great Britain because Americans simply did not produce the quality or quantity of cloth required to supply an army.[50]

Before the United States could become a military rival to Great Britain, it had to improve production. Supplies mattered not only for military performance, but also as recruitment tools, as most men enlisted for money—and clothing.[51] As manufacturers erected new wool and cotton mills throughout New England, the War Department charged Coxe with the collection of cloth supplies for uniform making. Secretary of War William Eustis instructed Coxe in May 1811 to purchase blue cloths for coats and woolen and cotton cloths for vests and pantaloons. The following February, Eustis wrote to Coxe again with clothing instructions, this time asking for all the blue cloth Coxe could procure, noting that some of the cloth could be drab or mixed color if need be.[52] Coxe had little knowledge about manufacturing outside the Philadelphia vicinity, so he relied on manufacturers to send him samples in response to handbills and advertisements, realizing that most material would have to come to the city for processing anyway.[53] Uniforms were made by tailors and seamstresses on a piece-rate basis—seasonally before the war and year-round following mobilization.[54]

New England newspapers, as they lamented Congress's injuring of commerce, printed news about the government's encouragement of domestic manufacturers. Many contained advertisements for American-made cloth for army uniforms.[55] Massachusetts manufacturers were hopeful. One congressman from the state had "been asked to submit to the Senate a resolution of the legislature of the Commonwealth of Massachusetts stating the perfect ability of the state to supply such blankets and clothing as the general government needs . . . as there can be no doubt of the Commonwealth to supply by contract any such articles mostly if not entirely from their own manufactories"[56] But while the War Department contracted directly with arms makers, it did not yet form these types of agreements with textile manufacturers. As early as 1809, manufacturers began sending Coxe cloth samples for inspection, hoping for contracts, or at least the promise of a new customer. Almy and Brown, seeking new sales opportunities in the changing economy, sent Tench Coxe cloth "for consideration" in March 1809.[57] Coxe continued to solicit materials. He sent a circular in July of 1810 to manufacturers throughout the northeast, imploring them to improve cotton drilling. He enclosed samples of Russian linen (raven's duck), which was better than twilled cotton drilling from Europe and America for light sails, tents, knapsacks, and wagon covers. He said that if manufacturers sent him suitable samples he would contract with them the following year.[58] They did not meet his expectations to an extent that would have enabled the United States to be self-sufficient in clothing production, as it was in arms.

War

When the conflict formally began on June 18, 1812, preparation had been under way for the previous five years, but finances and wartime bureaucracy were not yet in order. The embargo had decreased customs revenue. Congress, in expectation of a wartime jump in expenditure to $22 million, doubled the tariff and borrowed $11 million on March 4, 1812.[59] That same month, Congress replaced the purveyor's duties with three separate departments: quartermaster, commissary, and ordnance.[60] Lawmakers intended these wartime acts to shore up funds and increase organizational capacity, yet none created smooth wartime machinery. Interaction among the three sections of the War Department remained chaotic and the Quartermaster Department exhausted its funds by October 1812.[61] Americans, fearing the government's inability to make interest payments, had only subscribed to $600,000 of the federal loan by May 1812, prompting Congress to accept Secretary of Treasury Albert Gallatin's suggestion for an issuance of $5 million in treasury notes twelve days after the United States declared war on Britain. These notes, however, were not readily accepted, except by federal agents and by desperate creditors and contractors.[62] Partly for this reason, textile manufacturers could do better financially by producing for civilian markets, in contrast to arms contractors who relied almost solely on government demand. Yet, many who received contracts or felt encouraged to start a factory during the war were disappointed in its aftermath. The War of 1812 created demand for both textiles and arms, but it offered only fleeting opportunities for the majority of arms contractors and small textile manufacturers.[63]

Arms in War

The war required an unprecedented quantity of American-manufactured firearms. The federal arsenals had over two hundred thousand arms on hand, but not all of these were serviceable. Congress authorized the regular army to add thirty-five thousand men to their ranks, and although the numbers only ever reached about half that, the federal armories and private contractors struggled to supply the increased demand.[64] Of the eighty-five thousand stands of arms contracted for in accordance with the 1808 Militia Act, only 31,645 were delivered at the start of war.[65] Ohio Senator Thomas Worthington voted against the war specifically because he saw firsthand the lack of preparedness on the Northwest frontier.[66] The Springfield Armory responded to increased demand by almost tripling its production, from 12,020 muskets in 1811 to over thirty-five thousand the following year.[67] Harpers Ferry still only produced about ten thousand annually.[68] Because so many US-owned arms were outdated or damaged, the armory had to devote significant attention to repair in 1813. In 1814, Harpers Ferry dropped its output to only 9,585 arms, but it repaired 5,190.[69] Contract arms were slow in arriving and were generally inferior to those produced at the armories. Regardless, the government continued to rely on private arms makers to supplement the federal armories' output.[70] Throughout the war, the government cemented its relationships with a select group of manufacturers, while letting most of its nineteen contracts expire.

Just as in the 1790s, many arms manufacturers who had received contracts in 1809 and 1810 struggled to complete their government orders.[71] Most of these failures were smaller producers without the means or the political connections to withstand financial hardship. Struggling manufacturers had solicited the initial contracts because they desperately needed money, but as Chief of Ordnance Decius Wadsworth noted, "those who first engaged in the business at a later period, when higher prices were given, were more successful."[72] Unlike the 1790s contracts, post-1808 contracts offered cash advances, usually between 10 and 20 percent of the total cost stipulated in the contract. These payments were emblematic of the codependence between public and private resources. The government needed firearms; the manufacturer needed capital. Cash advances enabled private production but often posed problems for the War Department, because failures to produce represented sunk costs for the government. Rufus Perkins, for example, a manufacturer in eastern Massachusetts, received a cash advance of $2,687 for twenty-five hundred stands of arms, worth 10 percent of the $26,785 the government would pay upon completion of the contract. He never finished the order. As a small producer of arms mostly for individual militiamen and militia companies, Perkins saw opportunity in a large federal contract. Two years into his term, however, he had delivered only two hundred muskets. In 1813, the government terminated his first and only contract.[73]

Failures like Perkins's meant that the purveyor of supplies had to devote significant time to chasing after manufacturers and their sureties, which took time away from actually procuring supplies for the military.[74] When possible, it was better to contract with more established manufacturers to safeguard against default. Tench Coxe specified that he preferred to contract with "men of property."[75] Asa Waters was one such manufacturer—a prominent landowner and farm tools manufacturer in Millbury, Massachusetts, who founded an armory with his brother, Elijah, in 1808.[76] That year the brothers obtained a contract for five thousand muskets and a cash advance of $9,375.[77] Although new to the arms industry, the Waters brothers secured an advance worth over 13 percent of the total cost of the contract—an advance that represented a significant allocation of public resources to men with no prior federal government experience. It enabled Asa Waters to expand his factory. Several years later, he remarked that his workshops were fully employed, and that he could produce arms at an even faster rate if the government requested.[78]

The advances that manufacturers received varied widely. For example, William Henry of Pennsylvania received a contract for double the number of muskets as Waters, but an advance of only $100 more. This disparity in payments reflected the fact that in many ways the government, in addition to privileging producers in specific geographic locations, practiced what economic historian Joel Mokyr has described as "gentlemanly capitalism," in which one's reputation signaled his trustworthiness to fulfill market agreements.[79] As a purchaser of manufacturers' skills and capital goods, the federal government sought to minimize its exposure to risk, as well as maintain its network of valuable relationships. Part of this minimization of risk related to skill. During the execution of his first contract, Eli Whitney adopted die-forging and jig-filing processes (a die is a tool that cuts and shapes material, and a jig holds a piece of work in place), French techniques in metal-shaping that American arms makers sought to emulate.[80] More important than Whitney's technical abilities, however, was his social capital. It had taken him over eleven years to complete his contract from 1798, and some of his deliveries were of poor quality.[81] Despite the delays during his first contract, on July 18, 1812, Whitney signed his second contract for fifteen thousand muskets, which included federal reimbursements for factory construction.[82] Whitney knew federal officials personally and they in turn knew the men who served as sureties on his government bond. Whitney was a longtime friend of Decius Wadsworth, a fellow Yale graduate, who was appointed commissary general of ordnance on July 2, 1812. Whitney also knew Oliver Wolcott, former secretary of treasury, who served as the surety on Whitney's contract with the government.[83] When Callendar Irvine, serving as commissary general of purchases, terminated Whitney's contract in 1813 for an alleged failure to complete one of its terms, Wolcott petitioned Secretary of War John Armstrong on Whitney's behalf, and Armstrong ordered Irvine to reinstate the contract, which Whitney satisfactorily carried out.[84] Even though Whitney experienced confrontation with federal officials, just like William Henry had, and even though he was no more a skilled arms maker, Whitney's personal connections and proximity to Connecticut Valley production assured his status as a regular government contractor. The government's contract with Eli Whitney was emblematic of the mutual dependence of reputable private arms makers and the federal government during the War. Whitney's first government contract "saved him from ruin," but now the government would come to depend on him.[85]

The federal government also depended on Simeon North of Middletown, Connecticut, who became a successful arms manufacturer with the help of public patronage. Like Whitney, North was a property owner and had wealthy friends who endorsed his bank notes and served as his sureties.[86] After securing his second contract, North invested $100,000 in a new factory dedicated solely to arms manufacturing on Staddle Hill in Middletown, Connecticut.[87] Regular government support enabled him to stop all other manufacturing pursuits.[88] Soon after the war started, Secretary of War William Eustis visited North's armory and asked him to increase the size of the factory, whose water power made possible heavy machining beyond the capabilities of the waterworks at the Springfield Armory. The War Department rewarded North with a contract for twenty thousand pistols, the largest ever given.[89]

North's relationship with the government also reflected skill and compliance with contract terms. During the war, North traveled to Washington and exhibited several identical sample pistols to ordnance officials. As a result of this impressive display, North's 1813 contract, which was the first US contract to refer to the principle of uniformity, became a major stepping-stone in the development of interchangeable manufacturing. The contract did not use the word "interchangeability," but it contained the specification that any part of any gun should fit any other.[90]

True interchangeability required that all public and private armories machine-produce identical gun parts, which required not just innovation from manufacturers, but foresight on the part of federal bureaucrats. Historian Ken Alder has shown how the French state sought to assume greater control over the processes of production from private gunsmiths through its artillerists' management of the tools and skills associated with interchangeability. The French gun was, according to Alder, an "artifact essential to the authority of the state."[91] It became a priority in the United States during the War of 1812, at the same time the strengthening of government and military power became increasingly important.[92]

Ordnance officials were starting to recognize that uniformity in arms production mattered for battlefield operations. When hammer springs on muskets, for example, were not standardized, they could not "produce the most fire" and had "a great effect in deranging the pointing of the musket or the accuracy of fire."[93] Although the United States now produced a sufficient number of firearms, irregularity in the weapons given to soldiers during the war compromised soldiers' ability to change and repair parts. Reports from federal arsenals lambasted the poor quality of American arms. Many stands had been "overlooked by U.S. Armory inspectors" and came with "bayonets too weak in the neck that cracked a little, and several locks which pulled off too uncommonly hard." One inspector of arms for the Connecticut militia "could not pass them out," but advocated that the muskets would be "good and sound" if "cut down to thirty-nine inches."[94]

Following the war, ordnance officials asked the superintendent at the Springfield Armory to determine the "best means to be devised and adopted for bringing the manufacture of arms to a uniform standard and pattern in all of their parts." They requested that "muskets given out as patterns from the armories be strictly alike . . . in order that the conditions of the contracts now entered into by this department be made conformably thereto."[95] The Ordnance Department developed a system of inspection that required that all contract arms be examined by a government proof-master, who verified that the weapon matched the model provided by the Springfield Armory.[96] All "contractors' arms should be equal to those produced at the national armories," and inspectors should be "particularly careful and rigid in their inspections."[97] This regulatory system linked factory owners, federal inspectors and officials, and itinerant mechanics, who spread news of technological developments from North's factory in Middletown to Eli Whitney's factory in New Haven and Lemuel Pomeroy's Pittsfield factory, as well as to the Springfield Armory, and even down to Harpers Ferry. Colonel George Bomford of the Ordnance Department sent the superintendents of Springfield and Harpers Ferry to North's factory to observe his improvements in manufacturing pistols and then implement the changes at the federal armories.[98] They incorporated North's improvements into new patterns, which they disseminated to all their contractors.[99] In the second half of the 1810s, government stamps on arms became distinct marks of quality.[100] While patterns did not lead to perfection, the statistics were compelling: by 1818, under improved inspection methods, the New England arms industry had achieved a passing rate of 90 percent for the firearms manufactured at the armory, and 75 to 80 percent for those produced by individual contractors.[101]

Textiles in War

As the War Department nurtured a relationship of mutual dependence between the arms industry and the federal government, it took a more haphazard approach to the textile industry. On July 6, 1812, the United States declared it illegal to trade with Great Britain. Aspiring business owners saw profits in the void left by British broadcloths, drab cloths, cassimeres, and muslins, fabrics that US manufacturers could emulate reasonably well.[102] It was relatively easy to start a textile mill, especially compared to an arms manufactory, and as soon as domestic sales seemed imminent, many manufacturers built new factories.[103] Moses Arnold, William Bowen, Thomas Hubbard, and Benjamin Duick, for example, pulled together $5,000 in 1814 to expand their general-store operations in Woodstock, Connecticut, to include a cotton spinning mill.[104] In New England, well over twenty-five new companies started during the war, and others, already in operation, reported flourishing.[105] Although US involvement in the War of 1812 did not create a textile industry capable of clothing all of the armed forces, as it did the arms industry, war with Britain helped secure the existence of domestic textile manufacturing.

Even with the postwar economic downturn, war was imperative for creating the incentive for textiles to take off when they did.

Coxe had tried to encourage Americans to manufacture cloth suitable for the military. Before the war, he had sent samples of Russian duck and drillings (heavy linen cloth), requesting that manufacturers improve their wares to match foreign quality, and some had sent back samples in return. Amos Binney, the commander of the Charlestown Navy Yard, had also distributed circulars to textile manufacturers requesting samples of American duck.[106] But Binney complained of the quality, and, as late as 1819 Commissary General of Purchases Callendar Irvine lamented that it was nearly impossible to clothe the army solely with domestics because of high cost and low quality.[107] Despite government ads that ran for weeks at a time requesting such goods as woolen kersey for overalls and jackets, many manufacturers chose to produce goods that ordinary consumers would purchase, such as nankeens and ginghams.[108] Manufacturers selected these pale yellow, fine cotton cloths and lovely, checked woven fabrics for their success in both urban and western markets, even as they offered little in the way of military suiting. Samuel Slater and Bela Tiffany, for example, opened a textile factory in Oxford, Massachusetts, in 1812, choosing to specialize in finer specialized cloths like ginghams, nankeens, stripes, shirtings, and twists, none of which were the drab cloths, heavy linens, and kerseys necessary for military uniforms. Although New England markets suffered following the embargo, western demand was steadily increasing in places like Pittsburgh, Cincinnati, and Zanesville, Ohio.[109] The opening of the National Road in 1811 connected the Potomac and Ohio Rivers and facilitated the westward flow of goods and people to new cities. Slater and Tiffany, like many others, consigned their goods to Philadelphia, where they could be sold in urban markets or shipped farther west.[110] Almy and Brown, likewise, sent their goods to Philadelphia and the interior. They had sent samples to Coxe in 1809 when they needed new customers, but during the war they passed up government contracts. Even with high shipping costs and the uncertainty of sales, larger established firms preferred the consumer market over the government because the potential for high profits was greater. And when there was "risque too great," Almy and Brown simply instructed their agent to raise the prices.[111] Alternately, when "ready sales for cash" were hard to come by, owing to "so many cotton mills . . . and cloth so easy to be obtained," the firm directed the agent to "take to the interior" and sell the cloths at a 10 to 15 percent discount so that they would net about 20 percent profit on the invoice price after deducting commission.[112] Manufacturing establishments like Almy and Brown could afford to take risks in the flooded consumer marketplace.

It was not, however, unheard of for textile manufacturers to devote production to government orders. Several Philadelphia-area manufacturers became the nation's first textile contractors during the embargo. In addition to making such fancy goods as sophisticated woolen shawls, lace, and European-style carpets, Philadelphia manufacturers had been producing coarse woolens and linens for the market since the 1760s.[113] One such wool manufacturer was Samuel Winpenny, who entered into the first-ever contract for blankets for the US government in 1808. Another, James Kershaw, commenced factory operations just in time to become a government supplier. This marked a change from just several years earlier when military-grade goods were purchased from Philadelphia importers, rather than domestic producers.[114] Winpenny's son would continue business with the government in the 1820s and 1830s, and Kershaw remarked years later that he was unqualified to comment on the "general market" because he had "always been employed in making goods for the army alone."[115] Kershaw and Winpenny were unique, however, in their continuation as loyal public contractors after the war.

Some factories produced a combination of non-military wares and government contract items, which put them on shakier footing. Huntington & Backus Woolen Manufactory in Norwich, Connecticut, for example, manufactured both cheap satinets, which were sold to lower-class consumers who could not afford an all-wool fabric, and coarse woolen kerseys, which were sent to the military depot at Philadelphia, where some of the over three thousand tailors employed by the government would transform the coarse woolen cloth into coats and trousers for soldiers.[116] The owners had invested $70,000 in the company and were rewarded with handsome annual sales of $40,000 during the war.[117] When peace resumed, they ceased military production. Similarly, J. and S. Hindsdill's Woolen Factory in Bennington, Vermont, though significantly smaller than Huntington & Backus, produced for both military and civilian markets. During the first year of war, most of the factory's goods went to civilians, but in 1813 and 1814, the government provided the bulk of its income. In 1812, only $730 of the firm's revenue came from the one thousand yards of kersey it manufactured for the United States, while it received $10,375 for the twenty-eight hundred yards of fulled cloth (pressed woolens) it manufactured for general consumers. The following year, however, the Hindsdills produced twenty times as many kerseys, and reduced their fulled cloth output to two thousand yards. This shift in production likely resulted from price fluctuations. Fulled cloth sold for over $3.50 per yard in 1812, but only $1.50 in 1813. Conversely, the government raised prices to entice contractors; the rate for white kersey increased from 73 cents per yard in 1813 to 94 cents per yard the following year. The firm added blankets to its repertoire in 1814; they sold at $4 each and netted $9,600. When the war ended, both J. and S. Hindsdill and Huntington & Backus suffered. Military production stopped and the reintroduction of foreign goods into American markets crowded out domestic products. Hindsdill cut production almost in half during the first year of peace; in 1816, it cut it in half again.

For decades after the War of 1812, no blankets were manufactured in New England.[118] We might attribute this to the region's relative distance from Philadelphia (which remained the depot for military goods even during peacetime), but even there, for example, Globe Mill manufactured heavy printed woolens for the government during the war, only to be almost ruined when the postwar resumption of imports eliminated government demand.[119] Instead, it was a combination of the willingness to commit to government production, proximity to supply depots, and chance that determined a company's relationship with the federal government.

Other companies did not pursue military contracts. The same year that the Hindsdills ramped up production for the government, Patrick Tracy Jackson, James Lloyd, Francis Cabot Lowell, Nathan Appleton, John Gore, and Israel Thorndike incorporated the Boston Manufacturing Company in Waltham, Massachusetts, to produce brown and bleached shirtings and sheetings. Bleak commercial prospects prompted these merchants to decide that it made sense in 1813 to invest some of their wealth in textiles. Francis Cabot Lowell in fact had recently returned from a factory tour of England. Lowell is usually credited with developing the first truly American system of manufacturing; the system's origins, however, are anything but patriotic.[120] He was in Britain before and during the first year of the War of 1812, buying fine linens and other British goods while his home nation practiced nonimportation.[121] He also engaged in industrial espionage. The British state guarded national technology, which helps explain why Lowell's detailed record of his travels overseas reveals almost nothing about his factory observations. As Lowell traveled throughout the country with his family, he wrote only that an acquaintance named William Smith had given him a tour of the town manufactories in Glasgow and that he found "the manufacturing towns very dirty."[122] Nathan Appleton's nostalgic account of the introduction of the power loom referred briefly to the fact that Lowell and Appleton had seen each other in Edinburgh, where Lowell informed Appleton that he planned to visit Manchester "for the purpose of obtaining all possible information on the subject, with a view to the improved manufacture in the United States."[123] It is generally believed that Lowell committed much of what he observed to memory and then returned home to profit from his observations.

Capitalized at $400,000, the Boston Manufacturing Company was worth over five times the value of Huntington & Backus, and over fifty times that of J. and S. Hindsdill.[124] The company did not need a government contract, because its directors had every other advantage working in their favor. In fact, Callendar Irvine remarked that during the war the "large establishments . . . would not make a yard of cloth for our suffering troops."[125] The directors of the Boston Manufacturing Company maintained a range of investments in foreign trade, insurance, and banking, and were well-connected to legislators and consular agents all over the world.[126] During the war, for example, Francis Cabot Lowell received a letter from his selling agent in England acknowledging that if war continued Lowell could not send corn or flour to England, but that the government would issue licenses for him to take it to Spain or Portugal.[127] Others, like Nathan Appleton, were engaged in smuggling during the embargo and war, and because of their connections, had legal recourse to compensation for shipping losses incurred from illegal trade.[128] While other textile companies complained of a "want of capital" in the 1810s, the directors of the Boston Manufacturing Company had an easy time securing loans because they controlled stock in the major Boston banks.[129] This wealth and access to power made it possible for the company to buy up more land following the war when other companies struggled.[130] The same year that Hindsdill cut production in half, the Boston Manufacturing Company authorized its treasurer to contract for the construction of new buildings.[131]

Peace

On December 24, 1814, a treaty of peace and amity between Britain and the United States was signed in Ghent and ratified by Congress two months later. From the point of view of diplomats, nothing really changed. Boundaries stayed for the most part the same, the contentious issue of impressment (forced naval service) remained unsolved, and arrangements were made for the return of prisoners and slaves. But the treaty marked Britain's official acceptance of the United States as a sovereign nation with the prerogative to engage in commerce and expand its borders. It also highlighted the importance of military preparedness. In his message to Congress in February 1815, President James Madison urged Americans to remain vigilant despite the "pacific dispositions of the American people and the pacific character of their political institutions." While "resources were competent to the attainment of every national object," "preparation for war is not only indispensable to avert disasters, but affords also the best security for the commitment of peace."[132] For this reason, the federal government would maintain a regular army and would continue to provide contracts for arms. But "preparation" during peace only benefited some. After the War of 1812, the government renewed just a handful of contracts, and American manufacturers in general faced competition from the recommencement of foreign trade. Even before an armistice was declared, New England sellers feared the imminent fall of prices and the end of military requests.[133]

Madison's celebration of the resumption of commerce with Britain in his post-treaty message to Congress ignored the realities of the impending flood of British imports. His message certainly would not have resonated with Frederick Wolcott, a textile manufacturer in Torrington, Connecticut, who had built his factory during the war. Business prospects had looked promising, but by the time Wolcott had cloth ready for sale, peace had been declared and the large importation of British textiles drastically reduced the prices of his wares.[134] The owners of a small cotton factory in Coventry, Rhode Island, experienced similar commercial disillusionment following the war. They, too, had commenced operations at the end of the war, hoping for good profits, but instead faced a surge of imports. As a result, they decided not to add more spindles.[135] Even those who had government contracts could not count on its support. Huntington & Backus had contracted with the government for blankets and kerseys during the war, but they were left with the same postwar reality as non-contractors. They, like others, bemoaned the "flood of foreign goods that left them unable to dispose of their goods."[136] The directors of the Boston Manufacturing Company, though, had little to lament. The government's decision to close American markets to imports and ban US ships from the carrying trade had provided the space for their foray into textile manufacturing; its decision to declare peace resumed their trade with Britain. These men imported and sold the British textiles that made so many other manufacturers irate.

As US urban markets were flooded with "cheap" cottons and woolens from Britain, particularly half-finished coats that had been fashioned for military use during the Napoleonic Wars, wartime opportunities fizzled out.[137] Rhode Island mill owner Samuel Ogden believed that the postwar slump would be short-lived "because it is the effect of a sudden surprise and will turn more to the advancement of business."[138] He was wrong. Manufacturers in Amesbury, Massachusetts, unsuccessfully implored naval agents for government contracts; five years later Boston Manufacturing Company stockholders would purchase one of the town's failed woolen mills.[139] Some weapon producers were also hit hard by the Treaty of Ghent. Sweet, Jenks and Sons, for example, the lone gunsmithing operation in Rhode Island, had received a contract for four thousand stands of arms prior to the war, but in the absence of federal demand they devoted factory production to cotton goods.[140]

The manufacturers least affected by peace with Britain were the arms makers who had secured their status as regular government contractors. Federal production sites were still more efficient than private factories, but at the close of 1814, fewer than twenty thousand stands of arms remained in the federal arsenals, down from two hundred thousand at the start of the war, and War Department officials claimed that had the war continued, the country would have struggled to continue fighting with this shortage of arms.[141] Secretary of War John C. Calhoun argued that the federal armories could make up for this shortage by manufacturing higher quality firearms at lower costs than contractors; stipulations in the 1808 Militia Act, however, required that the federal government purchase arms from private producers.[142] Soon after the post–Militia Act five-year contracts expired, the Ordnance Department assumed administration of contracts, which had existed under the purview of a changing cadre of federal officers during the war. Decius Wadsworth, as chief of ordnance, would now be in charge of renewing contracts, and the relationships he had established with certain manufacturers, along with some contractors' reputations for reliability and quality, determined who received peacetime contracts.[143]

Following the war, the main contractors for the next two decades would be Robert Johnson, Nathan Starr, and Simeon North in Middletown, Connecticut, Eli Whitney in New Haven, Connecticut, Asa Waters in Millbury, Massachusetts, Lemuel Pomeroy in Pittsfield, Massachusetts, and Marine T. Wickham, the only non–New Englander. Wickham, who had worked as master armorer at Harpers Ferry until 1811 and federal inspector of contract arms at Philadelphia during the war, secured a contract in 1816 to produce four thousand muskets at $14 per stand annually for the federal government at his manufactory in Philadelphia.[144] This financial arrangement marked a vast improvement over Wickham's humble origins as an apprentice to a country gunsmith in Maryland. While cotton manufacturers complained of the ebbs and flows of business, these men's factories remained in "constant operation."[145] Although they depended on a strong economy for payment, the government was a guaranteed customer. Nathan Starr, for example, went to Washington to solicit additional funding and made note that the government would not issue additional treasury notes until trade duties started rolling in.[146] Nevertheless, he received his pay. For some, the military offered dependable business, even during times of peace.

Also, the United States was never truly at peace. Despite peace treaties with Indian tribes following the war, the government maintained a military presence on its frontiers, especially along the Florida border, where hostilities between the United States and Creek and Seminole Indians continued after the war.[147] Additionally, in the wake of war with Britain, members of the government recognized that American liberties had been won through armed force and that they could be threatened at any time by despotic Europeans.[148] A productive arms industry provided the antidote to real and perceived threats to national security, and the War of 1812 had spurred federal officials to strengthen the federal armories and the contract system. The development of a textile industry was less imperative for war-waging and national defense, because the government could and did import clothing for the military, but war had done much to foster the industry's growth. Indeed, the federal government began considering "the expediency of making provision by law for clothing of the army of the United States in domestic manufactures."[149] This began to seem possible because wartime trade policies and government initiatives had increased the prevalence of American cottons and woolens in domestic markets and on military supply lists.[150] By 1820 all of the army's cotton drilling, grey cloth, and kersey, the principle article of wool for military clothing, were domestic (only blue cloths, blankets, and several scarlet cloths were imported). American kerseys, in fact, had reached a point of "high perfection."[151] Following the war, the militia, unlike the regular army, were expected to supply their own clothing, for which they would be reimbursed. The fact that they were able to purchase domestically was a sign that the textile industry had improved.[152]

To a large extent, the federal state and its war-making capabilities had made this growth and improvement possible by engaging in this early form of national security capitalism. The government had provided the contracts, picked the winners, and established policies that shored up resources and spawned industrial development. A small cohort of contractors, along with the federal armories, would guarantee an annual output of arsenal capable of arming the nation, while a growing number of textile firms, of varying size and with different relationships to the government, would slowly but surely overtake foreign competition to clothe both civilians and armed forces. In 1800, American industry had seemed incapable of supplying an entire army, but in 1815, the United States emerged from a war with a major world power in which its soldiers had fought with American guns in their hands and at least some domestic textiles on their backs. And this was only just the beginning. Despite a postwar contraction, American industry would continue to expand and, with it, the nation.

Florida. From John Melish, J. Vallance, and Henry Schenck Tanner, *Map of the United States with the Contiguous British and Spanish Possessions* (Philadelphia: John Melish, 1820).
Courtesy, The Library Company of Philadelphia

[4]

Financing Industry through Florida

TWENTY YEARS AFTER the United States purportedly gained Florida in the 1803 Louisiana Purchase, the peninsula provided the venture capital for New England industry. After the War of 1812 had settled border issues with Britain and largely subdued the Delaware, Shawnee, Cherokee, and Creek Indians, the United States turned its attention to its Indian and Spanish foes in Florida, which Spain had refused to cede with Louisiana.[1] As cotton production pushed farther south, the acquisition of Florida in 1821 promised aspiring planters new opportunities for profitable landholding and guaranteed commercial access to the Gulf of Mexico for the export of southern produce. The expansion of slave country, as historian Adam Rothman has termed it, also provided a boon to New England manufacturing, not just in the form of increased supply of high-quality long-staple cotton for northern manufacturers, but also in the form of arms appropriations and financial rewards for men who were just beginning to transition from mercantile to manufacturing pursuits.[2]

In the decades between the purchase of the Louisiana territory and the coining of the phrase, "Manifest Destiny," American citizens and their government cooperated to expand and secure the nation's borders.[3] In the context of the contest for Florida, the US federal government engaged in targeted stimulus spending to expand its industrial resources in the North.[4] The terms of the treaty by which the United States acquired Florida from Spain provided remuneration to a select group of merchant-industrialists who had outstanding shipping claims against the Spanish government. These claims payments were then invested in northern factories. In an effort to shore up its control of the territory, the federal government pumped money and resources into its arms industry.

One of the executive's most crucial decisions in stimulating industry was to negotiate for Florida and in the process assume the mercantile debts of New England capitalists. The Transcontinental Treaty would transfer over $1 million in federal funds to some of the Boston Manufacturing Company proprietors as part of the United States' agreement to assume American citizens' claims against the Spanish government in exchange for the acquisition of Florida.[5] As the Panic of 1819 created capital shortages for many businessmen, these claims payouts coincided perfectly with large-scale expansion at Lowell, the nation's first industrial city.[6] They provided financial assistance above and beyond what the initially minimal protective duties offered.

We do not tend to associate government stimulus with early US economic development, but although the state's role is much more visible in early modern Europe or during the twentieth century, remuneration from the government subsidized new textile undertakings, just as contracts subsidized arms makers. By 1820 federal reimbursements for claims against the government were nothing new, nor was the supply of a standing army. What was new was the imposition of an unprecedented level of federal oversight on its arms production to ensure adequate supply and the assumption that the government would bear the responsibility for failed private commercial ventures. Targeted federal spending—part and parcel of the economic and military security goals of an expanding nation—made possible the expansion of industry in New England.

In the years between the Louisiana Purchase and the Mexican-American War, continental expansion should be understood primarily as the strategic consolidation of land on which to establish a sound political economy.[7] Financial expenditure in the service of armed force is usually associated with eighteenth- and nineteenth-century European nation-states, but the early American republic was no stranger to it. Public expenditures for territorial acquisition and defense became commonplace as the United States required ever more land to overcome constraints on its economic growth. Expanded markets for agricultural crops and manufactured goods formed the early republican solution to the problem of commercial fluctuations.[8] As historian Drew McCoy has shown, the Jefferson Administration negotiated the Louisiana Purchase with France because the territory promised to solve a key problem for Jeffersonian political economy. The North American continent offered seemingly endless land on which to hinge individual fortunes and national welfare. Historian Edward Baptist has labeled this territory "ghost acres"—extra land that allowed the northeastern United States to avoid Malthusian catastrophe.[9] Seemingly limitless land would enable the nation's population to spread out, operate independent farms, and dispose of surplus produce.

The Louisiana Purchase scarcely gave the United States meaningful control over "Louisiana." Even after treaties were signed and ratified, newly acquired territory required defense, which meant that western fortifications and Indian warfare began to absorb a significant proportion of public resources.[10] The signing of the Treaty of Ghent in 1814 by no means ushered in a period of uncontested peace; disagreements with Great Britain over commercial rights persisted and the potential for British naval intervention posed a particular threat to US military and commercial security. For this reason, Florida was of great strategic importance to the United States. Americans could not countenance another nation's control of a peninsula whose extensive coastline offered coveted commercial access to Caribbean and Atlantic markets. In foreign hands, the region threatened national security, as when the Spanish made it a haven for runaway slaves from nearby states.[11] Lax policing by Spanish officials made southern slaveholders especially fearful of another slave rebellion like the one that had occurred in Louisiana in 1811.[12] Commercial and military motives drove Americans to spend the subsequent three decades completing their aggressive pursuit of Florida—in essence, consolidating control over land nominally possessed.[13]

While the acquisition of Florida would seem mostly to have benefited white southerners, who sought new land for plantation ownership and the eradication of slave-tolerant Spanish rule, it boosted the business prospects of some New England factory owners. The road to economic development in fact traveled through Florida. If we step back from the local history of places like Lowell and view New England industrialization from Florida, the advent of New England manufacturing changes from one shaped by entrepreneurs and laborers to one shaped by geopolitical developments and federal patronage. The acquisition of Florida necessitated a steady supply of arms, as well as financial remuneration for the investors in what would become the greatest concentration of industrial resources in the United States: the Waltham-Lowell mill complex. The War Department implemented improvements in the management and production of arms supply in the context of the ongoing military demands but diminished military urgency, in light of peace with Britain. Expansion also provided a financial stimulus to textile manufacturing in the wake of the economic downturn of 1819. As the United States waged war against Seminole Indians and negotiated with Spain for Florida, it vastly expanded its manufacturing capability in a region a thousand miles away.[14]

In the years following the War of 1812, individuals in the War Department engaged in military state building, an administrative process of achieving martial preparedness. At the same time, industrial business underwent significant modernization.[15] Historian Alfred Chandler referred to these developments as the precursors to the post-1840s revolution in American business in which the "visible hand" of the manager, not the "invisible hand" of the market, dictated a firm's functions.[16] When we shift our focus away from owners, agents, and managers, we see the hand of the federal government—clearly visible in the case of arms production, perhaps more gloved in the case of the textile industry. This hand was influential in ways to which scholars have not always paid attention, even as they, for example, have debated the importance of tariffs. Tariffs were but one form of fiscal support among a host of policy options. Tariffs on muskets and pistols had been in place since 1789 (with pauses from 1794–1796 and 1797–1801) as a nod to both the importance of domestic manufacturing and the hopeful certainty that the United States would not need to rely on imports to arm its troops; it was bureaucratic changes, however, coupled with a financially assertive War Department, that prompted unprecedented productive capacity in the arms industry.[17] Textiles did not receive serious tariff support until 1816, and even then continual congressional debates about tariffs bred uncertainty on the part of manufacturers. Executive decisions about land and security, on the other hand, generated industrial growth and efficiency. The ramifications of these decisions are where we should cast our attention.

Textiles

In 1821, the same year the United States formally acquired Florida, Joshua Aubin's business associates sent him to Amesbury, Massachusetts, to purchase a woolen mill. There was no need even to break ground in Amesbury because a mill, with its old machinery, was already there. It had most likely been one of the casualties of the recent economic downturn and was "worth but little." Aubin, in fact, paid only $7,000 for it. The death of one man's business venture, though, was an opportunity for Amos Adams Lawrence and his business associates, who, back in Boston, incorporated the undertaking as the Amesbury Flannel Manufacturing Company, capitalized at $100,000.[18] This was only a preview of what was to come: one year after Aubin purchased the woolen mill on behalf of Lawrence and company, this same group of men applied for and received an act of incorporation for the Merrimack Manufacturing Company in Lowell.[19] Valued at $600,000, it was worth far more than any other manufacturing venture in the country. Most of its directors were members of the Boston Associates, a loose network of mercantile, manufacturing, and banking investors that included men from such illustrious New England families as the Lowells, Lawrences, Cabots, Appletons, Perkinses, and Otises.[20]

In the case of the textile industry, public capital followed private action. The directors of the Boston Manufacturing Company had established a textile manufactory in 1813 that presaged the success of mechanized domestic production and whose success soon was known internationally.[21] They, unlike many others, escaped the economic downturn of 1819 unscathed. British imports, price deflation, and general economic depression in some areas of New England drastically decreased textile output and led some producers to cease production altogether.[22] The 1820 census was replete with manufactures' lamenting the downturn and pleading for government help. In the "general remarks" section of the census questionnaire, the agent for the Winthrop Cotton Factory in Kennebunk, Maine, for example, wrote that the stockholders had planned to stop operations the previous winter because things were so bad. Relying on what they "supposed and confidently expected Congress would do for Manufactures," they made an additional expenditure of $1,000 that spring, but greater tariff protection was still several years away and sales remained low.[23] The Boston Associates, on the other hand, had good reasons to expect government support because, after decades of successful (and sometimes not so successful) international commercial ventures, they received public capital for failed shipping ventures off the coast of Florida.

If we follow the money, we can trace the personal and political networks that gave certain New England industrialists access to a huge subsidy at precisely the moment of seemingly fortuitous industrial takeoff. The development of large-scale manufacturing in New England then becomes less the result of individual ingenuity and personal wealth and more the result of expansionist diplomacy and political wrangling.

When Secretary of State John Quincy Adams negotiated with the Spanish minister Luis de Onís to acquire the Floridas and establish the United States' southern boundary west to the Pacific coast, he agreed that the United States would assume $5 million of American citizens' claims against the Spanish government. These claims stemmed from Spanish land grants made to American settlers in Florida and from commercial losses sustained under Spanish jurisdiction.[24]

Along with disagreements over the Unites States' border with New Spain, which included present-day Mexico and most of North America west of the Mississippi River and south of Canada, a dispute over shipping damages had plagued US–Spanish relations for decades. Spanish officials had failed adequately to police naval violence in their waters—a negligence that prompted American merchants to establish long lists of claims against the Spanish government. Many shippers claimed that their cargoes were unlawfully seized at customs or that pirates ransacked their ships. Marine insurance companies from Boston, New York, Philadelphia, and Baltimore made up a large portion of the spoliations claimants. They pleaded indemnity for the cases of rum, sugar, and silks their subscribers had lost at the hands of privateers and pirates.[25] While the first sentence of Article IX of the Transcontinental Treaty mandated that the United States and Spain renounce all claims against each other, the US government alone offered to repay its own citizens for shipping damages caused by the other power's faulty management of its ports. Article XI pegged the extent of the US pledge at $5 million.[26] In the geographic breakdown of the claims' distribution, New England claimants received $1.8 million, Philadelphia $1.3 million, New York less than $1 million, Baltimore $600,000, and all areas south of the Potomac only $300,000 combined. A core group of Associates, including Thomas Handasyd Perkins, Nathan Appleton, and Henry Cabot, and their various insurance and trading companies, received over half of the claims paid to New Englanders.[27]

The claims payments the Associates received are important both because they reveal the allocation of federal power for business development and because they represent a source of industrial capital that has been neglected by scholars who point to the private commercial wealth that funded New England textile investments.[28] The claims payouts were financed by a government loan contracted in 1823 that was then paid for with taxpayer revenue via the 1824 tariff, which conveniently also represented the Associates' interests.[29] If we consider the claims payouts as subsidies, akin to the annual appropriations the federal armories received, the role of federal intervention in the development of large-scale textile manufacturing in New England becomes more obvious. With Articles IX and XI, Adams, long a supporter of industrialization and national development projects, built into the treaty the opportunity for financial gain for northern companies.[30] The $5 million that the United States agreed to pay Spain did not represent the sum total of all cases compiled by Americans against the Spanish government, but Adams did not want to subject the federal government to unlimited liability for American claims.[31] Onís, for his part, resented the fact that the United States did not pay more for its acquisition of Florida. The capped amount meant that not all claimants would be reimbursed. Soon after the conclusion of claims adjudication, the Senate Committee on Foreign Relations received a host of petitions claiming insufficient indemnity.[32]

In light of these charges of unfairness, it matters which claimants received rewards. In a letter to Congressman Joseph Hopkinson from Philadelphia, Congressman Daniel Webster, serving as the lawyer for the Associates, referred to the northern cases that he represented as "the *real commercial* losses, which occasioned the Treaty."[33] Webster's suggestion that claims by northern merchants provided Adams's impetus in negotiating the acquisition of Florida is perhaps an overstatement of the links between specific interest groups and diplomacy; the Associates, however, undoubtedly reaped the benefits of Article XI because of political connections and because of the nature of their business pursuits.

The expanded nation that John Quincy Adams helped build with the Transcontinental Treaty was constructed not just from the plantation system or from Indian land claimed by a European power, but from a series of investments, political deals, and connected favors.[34] It was not by chance that specific merchants, entrepreneurs, and investors received such favorable treatment from the US government. The same men who had founded a nationally celebrated textile manufacturing company also financed the national government through bond buying, and they were influential figures in Massachusetts politics and philanthropy.[35] Yet they had previously opposed federal policies that promoted expansion and economic protectionism. Some of these Boston investors were among the minority of Americans who opposed the Louisiana Purchase, and various members of this cohort had bypassed trade restrictions between 1807 and 1812 by shipping merchandise through Canada and Amelia Island, Florida. Some had even engaged illegally in the slave trade.[36] This illicit trade was one strategy to make the best of prohibitory commercial regulations; the founding of the Boston Manufacturing Company was another. The Boston Associates were rewarded for both. Although they previously opposed federal policies geared toward expansion and development, by the 1820s they reaped the benefits of a growing nation and national government.

The Associates were on familiar terms with the major players in the treaty settlement process. For starters, for the first time in almost two decades, they were on good terms with the president.[37] President James Monroe arrived in Boston on July 2, 1817, as part of his "good will tour" and met with members of the "Essex Junto," a group of Federalist party leaders, among them T. H. Perkins, who had three years earlier represented Massachusetts in the subversive Hartford Convention. Despite his Republican politics, Monroe established close ties with various Federalist merchants. In addition to maintaining a mutually rewarding relationship with northern trading enterprises, such as John Jacob Astor's American Fur Company, Monroe's administration granted special political privileges to some northern Federalists. After Monroe defeated New York merchant and politician Rufus King in the presidential election, he made King an ally of the administration. As part of this relationship, Secretary of State Adams sent King confidential briefings on the affairs of the state.[38] King kept regular correspondence with a number of Massachusetts politicians and merchants, including his brother William, who would serve on the Spanish Claims Commission, and US Senator from Massachusetts Harrison Gray Otis. During the year following the treaty's signing, Otis repeatedly urged the Senate to ratify it.[39] Otis, linked by family, politics, and money to the Associates, and an investor in the textile industry and shareholder in several Massachusetts insurance companies, understood that he and his Boston constituents stood to gain from the expansion of the United States into Spanish Florida.

The Boston Associates were also able to influence the commissioners' decision-making process. The Spanish Claims Commission, a three-member committee created by President Monroe in March 1821 with the advice of the Senate, proved essential in ensuring that the Associates benefited from the treaty. Webster, their legal representative, personally knew just one member of the commission—William King, a former Massachusetts politician who had recently become the first governor of Maine. King was no stranger to the Associates themselves either. He controlled several townships in the District of Maine with Senator Otis, who would serve as shareholder of several New England factories. Neither Webster nor the Associates had strong ties to the other two members of the commission; however, the commission sought advice from merchants on how to assess shipping claims, which worked to their advantage. Peter Chardon Brooks, Webster's most prominent client, provided much of this advice.[40] Brooks engaged in extensive correspondence with William King over how to determine merchants' reimbursements, and he instructed King on calculating the value of a ship's freight and insurance. He requested government aid in securing shipping documents from Spanish ports. Brooks also asked King, as a prominent resident of Maine, for a personal favor in locating the partners of an insurance firm in Bath, Maine, so that he could collect the papers related to the capture of one of his ships in 1797.[41] His correspondence with the commission proves that members of the Associates maintained ample ties to federal decision making. This worked out well for Brooks; the commission approved most of his claims.[42]

Daniel Webster's relationship with the Associates started at the outset of the 1815–1816 session of Congress, when Webster met with Francis Cabot Lowell in Washington. As a US representative from New Hampshire, a state with robust maritime interests, Webster had opposed tariffs, but after meeting with Lowell he modified his position to advocate a moderate cotton duty that would protect the Associates' nascent manufacturing company from foreign imports.[43] In the years before the treaty was signed, Webster also represented Otis, Appleton, and several of the Cabots and Lowells, and he helped broker the Associates' amicable relationship with the Republican administration, which no doubt helped during negotiations.[44] Webster, in fact, was on the payroll of the Boston Manufacturing Company for his regular legal services to the directors.[45] His relationship with the Associates explains why, during the Spanish Claims Commission, Webster indicated that some cases received better care than others. In a June 1822 letter to Boston broker Samuel Jaudon, Webster updated Jaudon on the state of cases and memorials and added that "some from Salem [whence Thorndike and others shipped their goods] you will observe to be well arranged."[46] Similarly, Webster sent a letter to Alexander Bliss, who ran the Boston law office while Webster was in Washington for the Claims Commission, instructing him to take whatever measures necessary to strengthen Perkins's case, even if the proper papers were missing."[47] Nathan Appleton, Peter Chardon Brooks, and Israel Thorndike all made claims for losses in the West Indies that had been postponed owing to lack of evidence, but Webster continued to work toward a settlement.[48] The contract prohibited Webster from representing their "losses [that] happened from smuggling, illicit trade, and the Slave trade"; however, he did his best to procure papers for even the more dubious cases.

Much of the Associates' West Indies trade involved the circumvention of US shipping laws, as well as the occasional transport of slaves.[49] Perkins, for example, engaged heavily in the West Indies trade, including its slave trade, after it was banned.[50] He would receive roughly $132,000 from the claims and would go on to invest over $100,000 in several textile factories and become president of the Appleton Company.[51] Israel Thorndike, one of the original backers of the Boston Manufacturing Company, would invest more in the various textile ventures than any other individual.[52]

Documentation of the Associates' investments does not reveal their precise origins, but it does suggest strong links to their Spanish claims income. Because many of these claims stemmed from insurance underwriting, the Boston Associates received additional payment for insurance claims they had long before settled locally.[53] Shareholders in companies such as Massachusetts Hospital Life, the Boston Marine Insurance Company, and the Massachusetts Fire and Marine Insurance Company, received additional returns on investment as a result of these claims.[54] This windfall occurred right as textile manufacturing was taking off.

For a group of businessmen who were concerned above all else with the safety of their wealth, the timing of the Transcontinental Treaty increased the soundness of production plans for their manufacturing companies.[55] The Associates knew by 1819 that they would receive federal funds, and the following year construction and production at Waltham increased rapidly. Several months after the ratification of the treaty, a group of Bostonians, including Peter Chardon Brooks and Thomas Handasyd Perkins, convened to commission Daniel Webster to represent them before the Spanish Claims Commission. In 1822 the proprietors of the Merrimack Manufacturing Company drew up its articles of association.[56] Soon after, they started the Taunton Manufacturing Company (1823) and the Hamilton Manufacturing Company (1825), prompting Adams to praise Perkins for his "truly patriotic undertakings."[57] These companies were followed by the Appleton and Lowell Companies in 1828 and the Suffolk, Tremont, and Lawrence Companies in 1830.[58] They also began buying up the debt of other manufacturing companies, just as did the insurance companies in which they owned stock.[59]

The public funds from the Transcontinental Treaty enabled the recipients to engage in the investment strategy historian Edward Balleisen has argued became increasingly common in the decades before the Civil War: vulture capitalism.[60] The shipping failures of the Boston Associates had been subsidized by the government, positioning them to take advantage of others' failures. The landscape created by the Panic of 1819 was ripe for this type of predatory purchasing. The Rockland Manufacturing Company in Scituate, Rhode Island, for example, had cost $7,000 to build and was capitalized at $23,400, but in 1819, it sold for only $750 at public auction.[61] When Joshua Aubin went to Amesbury to purchase a textile mill for the Lawrences, he entered a town that similarly had been hit hard by the economic downturn. The owners of one textile factory in the town had solicited the Navy Department several years earlier for patronage, writing that "the times are requiring a large capital" and "people in this town are out of employ."[62]

Failures like this throughout the region offered opportunities to those with capital. While the Associates usually chose to buy land and construct their own factories rather than purchase abandoned mills, they began purchasing what eventually became controlling stock in firms. The Dover Cotton Factory, chartered by the New Hampshire legislature in 1812, was reorganized in 1820 to raise additional capital by offering greater stock options.[63] The company's share offerings began attracting Boston investors in the early 1820s, and by 1827—reorganized again under the name Cocheco Manufacturing Company—members of the Associates and their insurance companies effectively controlled the factory through share ownership.[64] Additionally, the Boston Manufacturing Company owned stock in the Nashua and Amoskeag Companies, which they eventually controlled.[65] They also simply bought up large plots of land on which to expand their enterprises. For example, soon after its incorporation, the directors of the Merrimack Manufacturing Company made plans to build another factory, requiring additional land purchases.[66] In 1826, they received from the Boston Manufacturing Company two-thirds of the real estate the latter had recently purchased from two local farmers.[67] Nor did they limit themselves to these Merrimack River purchases.[68] In 1823, the agents of the Boston and Springfield Company bought up land and water rights on the Chicopee River in western Massachusetts from the Belcher Iron Works and incorporated the Chicopee Manufacturing Company.[69]

Predatory purchasing carried over into the realm of invention. Liquid capital enabled some companies and individuals to succeed in the federal patent market. The US patent system permitted the exchange of patent rights but left the adjudication of these rights up to individual courts. The Boston Manufacturing Company, which maintained at least one-third of its capital in liquid form, practiced a combination of patent infringement and purchasing to maximize the benefits it reaped from this system. The company kept a running account of patent payments and established a committee expressly to deal with the issue of patent purchases.[70] For those whose legal rights the firm violated, the only recourse was an expensive lawsuit, and so some who were victims of their predatory behavior just gave up completely. James Stimpson had patented two of the motions used in the Waltham looms—the feed and the application of the arms to one shaft for the lay motion—yet rather than assume the expense of pursuing justice through the court system he agreed to give up his rights for "less than what I consider I deserve."[71] The directors of large firms like the Boston Manufacturing Company had no trouble flouting others' patents, even as they were unsympathetic to those who supposedly infringed on their own.[72] Several years after Stimpson opted for minimal remuneration instead of pursuing a costly lawsuit, the Locks and Canal Company, a textile machinery company run by Boston Manufacturing Company stockholders, began using Aza Arnold's patented double speeder (an added motion to the roving of cotton). The Rhode Island machinist had patented his improvement in 1823, but this did not safeguard him against infringement. When he sued the biggest patent infringers, the Proprietors of the Locks and Canal Company, for $30,000, he was unable to prove their violation of his right. Arnold finally gave up when his patent ran out.[73]

This financial upper hand had been aided, in part, by federal reimbursements. While the claims were just one source of funding among many, they were nonetheless important in the context of federal "encouragement" of domestic manufacturing. The execution of Articles IX and XI of the Transcontinental Treaty represented federal intervention on behalf of large-scale manufacturing. These sorts of funds were inaccessible to most textile manufacturers, who desperately petitioned Congress for new tariff legislation in the midst of the 1819–1822 recession.[74] These manufacturers had to wait until the tariff of 1824 to receive greater protection for American products. And while the 1824 legislation was at least a nod to the importance of protecting American industry, its impact on textile manufacturers is debatable, and it paled in comparison to the cash payouts the Associates received.[75] Furthermore, the legislation skewed in favor of the Associates' companies, whose interests Daniel Webster represented. Unlike small companies, the Associates preferred limited protective duties, so as not to interfere with their own import trade. These minimal tariffs gave their goods just enough protection to compete against British imports, while being low enough to prevent the success of smaller domestic manufacturers.

That is not to say, though, that small producers were left without advocates. Philadelphia publisher Mathew Carey had long been a champion of protection for American industry, publishing works that promoted pro-industry policies and speaking publicly on manufacturers' behalf. In 1819 Carey chastised the US government for adopting a "ruinous policy, discarded by all the nations of Christendom, except Spain, Portugal and the United States" that denied sufficient tariff protection to domestic industry.[76] National decay was imminent, Carey argued the following year, if policy makers did not change course.[77] As the Spanish Claims Committee wrapped up its work, proprietors of small factories in Rhode Island—the ones who were luckier than the owners of the Rockland Manufacturing Company—sent letters to Rhode Island politician and manufacturer Zachariah Allen, who represented their interests to the Senate Committee for Commerce and Manufacturing. On their behalf, Allen requested an additional duty on imported woolen goods.[78] Historian Daniel Peart has shown how this type of lobbying and petition-making was an effective means of achieving favorable tariff policy in the early 1820s.[79] The resulting tariff, though, was modest: the ad valorem tariff increased from 20 to 25 percent.[80] By 1830, some woolen manufactures would finally receive heftier protective duties—in some cases a 50 percent tariff rate—but all of these policy fluctuations were bad for business.[81] A man known to the historical record as Mr. Winterbotham, for example, was forced to give up to his creditors his woolen manufactory on the Housatonic River in South Britain, Connecticut, in the late 1820s. He cited tariff uncertainty.[82]

Many companies who had languished during the depression survived despite the minimal protection for wool and cotton manufactures, but as the Associates' undertakings fanned out over New England, these smaller companies remained small.[83] Others never even made it through the depression. In 1819, Waltham Cotton and Wool Factory agreed to the $16,000 selling price the Boston Manufacturing Company offered it—$9,000 less than the Waltham Factory had proposed in 1817. At that time, the directors of the Boston Manufacturing Company had deemed it "not expedient" to purchase real estate. Two years later, however, when the economic downturn increased the direness of circumstances for the Waltham Factory and the stockholders of the Boston Manufacturing Company anticipated public capital from the treaty with Spain, vulture capitalism became expedient.[84]

Arms

The capitalists who received claims payouts owed their fortune indirectly to American gun manufacturers. The process of "acquiring" Florida required military power, both before and after the treaty with Spain. When Mathew Carey said that "the policy we have adopted renders us dependent for our prosperity on . . . the wars and famines of Europe," he referred to the United States' reliance on European circumstances that inadvertently benefited American manufacturers.[85] Carey was not wrong, but he might have acknowledged Americans' dependence on domestic warfare as well. The federal state's policing of borders for runaway slaves and "hostile" Indians hinged on and increased the business prospects of weapon manufacturers.[86] In the decades following the War of 1812, as Congress reduced the size of the military establishment, and military power seemed to be receding, individuals in the Ordnance Department made the most of their governmental license to increase and improve the nation's arms supply.

As historian Mark Wilson has contended—countering Ira Katznelson's arguments about legislative checks on the military that allowed the nineteenth-century United States to transcend European absolutism—it is foolish to think that the War Department was subordinate to Congress.[87] In the words of one congressman, "it would seem that the Executive branch of the government, together with the military, claims the right to make war at pleasure, without the sanction of Congress."[88] This capacity made possible, and *was made possible by*, the Seminole Wars.[89] Although these wars loomed larger in congressional appropriations, War Department writings, and horror stories of Indian massacres than they did in public discourse, they made a big difference to industry. Ordnance Department officials renewed contracts and shored up financial resources in an effort to provide adequate weapon supplies for an increasingly aggressive nation.

The 1810s and 1820s were key decades of military state-building. Never a popular institution, the military remained unpopular among American citizens following the War of 1812.[90] Congressmen's letters to their constituents reflected the pushback they received against any expansion of the military. They assured voters that any support they offered for the maintenance of an armed force resulted only from dire security concerns, and repeatedly presented arguments in Congress for further reducing the size of the army and the resources for the federal armories.[91] Although to many Americans the United States seemed to be at peace between the War of 1812 and the Mexican-American War, fighting on the frontier never ceased. The Seminole Wars in particular absorbed an incredible amount of resources from the 1810s through the 1850s.[92] Americans' ability to fight these wars—even though never officially declared as such—depended on weapons and the organizational capacity of federal bureaus.

The United States' struggles to equip its army during the War of 1812 had brought to the fore the importance of military logistics.[93] Chief of Ordnance Decius Wadsworth convinced Congress to expand the Ordnance Department because it was not equipped at the time to "discharge its duties."[94] The "Act for the better regulation of the Ordnance Department," passed in 1815, expanded the size of the Ordnance Department and granted the colonel of the department autonomy in establishing military depots, appointing master armors, orchestrating supply, and overseeing the inspection of all arms.[95]

Although War Department officials generally recognized that federal armories could manufacture arms of better quality than could private manufactures, the department continued to draw its supply from both the armory and private contractors.[96] This required adept management. Sociologists and entrepreneurship scholars have studied "institutional entrepreneurs," who use regulatory change to promote and disseminate new organizational practices.[97] Wadsworth was one such government entrepreneur who disrupted institutional norms to create industrial innovation and set precedents for ordnance officials who succeeded him. Prior to his tenure as head of the Ordnance Department, a coordinated set of standards for the federal armories and its arms contractors did not exist. He designed a timeline for achieving uniformity, requiring the national armories to produce pattern muskets by 1815 and begin full-scale production of the "Model 1816" the following year.[98] Although contractors like Eli Whitney and Simeon North had already begun working toward interchangeability, it was a slow and uneven process that required significant oversight and cost. Historians of technology have demonstrated that a factory needed to produce at least one thousand guns to make interchangeable parts production worthwhile.[99] In the early nineteenth century, only the federal government was able to devote the resources to this. It did so because of officials like Wadsworth, whose pursuit of interchangeability and increased production levels bolstered War Department military aims against congressional fiscal priorities and anti-militarism.[100] Wadsworth persistently "suspected an attempt at the next session of congress would be made to reduce rather than increase the national armories."[101] He spent the rest of his tenure working around that suspicion, and laying the groundwork for his successors to do the same.

On November 21, 1817, after a small band of Seminoles refused to vacate lands just north of the Florida border, General Edmund P. Gaines attacked an Indian settlement, and over the next two years American troops led by General Andrew Jackson waged war against the Seminoles in Spanish territory.[102] While Congress never formally declared war, both Secretary of War John C. Calhoun and President James Monroe advocated violence against the Seminoles for the sake of punishment and national interest, only later appealing to Congress that the military action was necessary for the "safety of our fellow-citizens."[103] Jackson's disregard for international law when he executed two British merchants who were selling guns to Seminoles by no means improved the reputation of the military in Congress, which continued to push for a reduction in its budget. The House of Representatives requested that the War Department present a report on the feasibility of reducing the army and called for "the attention and effort of the government to be roused to confine its excesses within the most moderate limits which may be practicable."[104]

In spite of petitions to reduce the military budget, the Ordnance Department assertively pushed arms production. As Wadsworth noted, "it would not be prudent to remain dependent on foreign countries for articles intimately connected with security. Measures ought to be taken . . . so we can wage a vigorous war. A great capital is needed."[105] Congress dictated the amount of money spent on arms production—$200,000 every year for arming the militia as stipulated by the 1808 act, varying annual amounts for the armories, and additional monies as military needs arose. War Department officials made routine presentations to the president and to members of Congress about increasing appropriations for arms production.[106] Sometimes they waited "until the temper of Congress has cooled" to ask for more money.[107] As skirmishes between Seminole Indians and Americans increased along the Florida border, Secretary of War John C. Calhoun hoped to "increase the product of the national armory." He told Wadsworth to authorize contracts for five thousand barrels and rods prior to receiving approval from Congress. Wadsworth, accustomed to circumventing the legislative hurdles, "expected [the plan for new contracts] to be fully understood at the next session of Congress and measures taken to make it effectual."[108] At another point, Wadsworth instructed the Springfield Armory to produce at least twelve thousand stands of arms, even if it meant exceeding congressional appropriations. The debt, Wadsworth said, could be made up later.[109]

Ordnance officials also engaged in clever accounting methods that made it appear as if less money had been spent. While every government agency required congressional approval of its finances, the Ordnance Department was unique in its development of sophisticated bookkeeping that enabled the department to present figures in a way that was more palatable to Congress. Beginning in 1816, the department required all officers of armories and arsenals to submit quarterly returns, which were then summarized for Congress.[110] Following directions from the Ordnance Department, the superintendent of the Springfield Armory, in preparing his statement of materials on hand at the armory, accounted only for an amount that would appear to fit within the prescribed budget.[111] He decided to include recent purchases for supplies like musket stocks on the next return.

Ordnance officials' financial activities sustained the technology districts they had formed during the War of 1812. When the War Department decided to contract for individual arms parts, such as barrels, bayonets, and ramrods, the Springfield Armory advertised for these weapon parts in local papers.[112] The labor transfer between and among the private and public sectors in the Connecticut River Valley that the War Department had cultivated paid off. Machinists shared technical knowledge when purchasing machine tools from other firms.[113] When the department needed supplies beyond what its employees produced, it could turn to others in this region to maintain adequate weapon production in the midst of the Seminole War. These additional contracts provided business to some general machinists who did not have the factory capacity to manufacture complete arms. Springfield Manufacturing Company in Ludlow, Massachusetts, for example, which existed primarily as a wool carding business, secured its first contract for the manufacture of five thousand musket barrels in 1817 and continued to make barrels through the 1820s.[114]

For contractors whose factories were devoted to federal production, public capital solved the problem of mechanical expense. The royalty fees for cutting-edge gun-stocking machines, for example, prevented Asa Waters from implementing them in his shop until 1825. Left to his own devices, Waters probably would not have purchased such machinery, but the machinery was necessary to comply with Ordnance Department specifications, and cash advances lessened the financial burden.[115] Sometimes contractors needed more than regular cash advances. Several years into his federal contract, North was in debt, despite government advances of $27,000. Wadsworth, ever the relentless advocate of interchangeability, convinced Secretary of War John C. Calhoun that the Ordnance Department should grant North whatever additional aid he requested.[116] Wadsworth understood the importance of supporting a manufacturer who was indispensable to arms innovation. He was not wrong. In 1816 North manufactured the first successful milling machine, a requisite machine for the system of interchangeable manufacture.[117] The device achieved a high degree of precision by mechanically feeding a table holding the work piece (or part to be cut, shaped, and smoothed) into a rotary multiple-toothed cutter. This machine represented, according to historian Merritt Roe Smith, "the first glimmerings of interchangeable production."[118] In 1824 the Springfield Armory implemented one of North's machines for milling bayonet muzzles.[119] North also eventually developed a model carbine, which gave mounted cavalry the dexterity of a shorter rifle and embodied the War Department's goals of uniformity and modernization.[120]

The experiences of men like North suggest that cash advances drove innovation. The early US patent system was notoriously difficult to profit from, even as increasing numbers of Americans sought patents for inventions.[121] North, for instance, did not patent any machinery until after 1840. Patent enforcement was difficult, especially prior to reforms in 1836, and so in many cases it was in the armorer's best interest to work diligently toward improvement and assume that his contract would be renewed. Whitney, for example, had experienced the limitations of patent protection with his cotton gin and never patented any arms machinery; he found government contracts a better inducement to innovate. Although Whitney never succeeded in complete interchangeable production, federal officials maintained that for anything Whitney pursued "in earnest" they could "expect as much uniformity as can be desired."[122] And so Whitney profited from federal contracts, while the US government reaped the subsequent mechanical benefits. The Springfield Armory implemented several of Whitney's inventions—a bridle and the mechanisms for forging metals and for driving a trip hammer by belt—without having to negotiate for their use.[123]

Once the federal government had invested public money in supporting regional producers, these manufacturers were not left to the whims of the market. Even during lean times, when War Department officials "complained of the small appropriations," they reassured Nathan Starr that he would receive federal payments.[124] It no doubt helped that Starr maintained the type of relationship with ordnance officer George Bomford in which the two could "have a good laugh" about the inferiority of Russian-made swords.[125] Robert Johnson, likewise, received a recommendation from Springfield Superintendent Roswell Lee for making rifles that he took to Washington, DC, in 1819 to secure a new contract.[126] Several years later, Johnson, North, and Starr traveled to Washington again to request more business.[127] Their requests were granted. Whitney, Pomeroy, and Waters also got new contracts.[128]

These men had satisfactorily completed their contracts throughout the 1810s and would continue to manufacture under government patronage despite administrative changes. By "An Act to Reduce and Fix the Military Peace Establishment of the United States," passed March 2, 1821, the Ordnance Department merged with the Corps of Artillery, losing its status as an independent bureau.[129] The economic downturn of 1819–1822, combined with a resentment of Wadsworth's and the department's disregard for federal budgets, prompted the demotion of the Ordnance Department as a means of cutting costs. The aggressive promotion of arms production initiated under Wadsworth, however, continued. In 1823, Representative Romulus Saunders of North Carolina noted that his examination of ordnance accounts over the past few years "showed that something like a check should be put upon the expenditure of that department," and that "some of the expenditures would strike this House with astonishment."[130]

However much Saunders and others wanted to "check" Ordnance, it was increasingly difficult to do so. When Wadsworth retired from the Ordnance Department, his successor in the demoted bureau, George Bomford, continued Wadsworth's precedent of shrewdly maximizing financial resources. After one of the main workshops at the Springfield Armory burned down, he pushed for additional funds to rebuild it.[131] When there were surplus funds, Bomford instructed superintendent Roswell Lee to spend them quickly, before they dried up.[132] Just as Wadsworth had engineered account statements to make them more palatable to Congress, so too did Bomford and his subordinates. One officer instructed Lee to fill out his expense form so that "it gives at the same time the means of accounting for any difference between the report and the paymaster's account."[133] Bomford was especially skilled at making the bureau seem simultaneously deferential to Congress and too financially and technologically sophisticated for lawmakers to understand. He instructed Lee to avoid debt so that when the armories needed credit Congress would be less likely to question their request. He also advised one of the ordnance captains to scale back plans for a new building at Springfield because he feared the presentation of the plans would invite a rejection of the whole proposal.[134] At the same time, however, Bomford presented expenditures in too abstruse a way for Congress to comprehend. In a postscript to one report to Congress, for instance, he wrote, "In some of those years, the expenditures and the products of the armory appear to be disproportionate. This has arisen partly by the purchase of a greater or a less quantity of stock and materials in a year than was consumed within it, and partly by the pay't of debts in one year, which had been contracted in the year preceding."[135] Perhaps the best expression of Ordnance's efficacy and Congress's resistance to it was Ohio Representative Joseph Vance's comment during congressional debates over Ordnance spending that "it had rather been necessary to hold a tight rein over those gentlemen, who are always sufficiently ready in devising ways and means to apply the resources of the country."[136]

Because Bomford successfully managed the bureau's relationship with Congress, armory superintendents knew they could rely on him to find them additional funds.[137] One of Bomford's subordinates assured them that "you will have the full amount allotted and if you desire it probably a few thousand more."[138] The Ordnance Department won its contest for autonomy, regaining its independence from the Corps of Artillery in 1832. Congress decided that the merger of the two bureaus "had been tried, and had not succeeded."[139] The ordnance bill passed the House 101 to 66, but not without a lengthy protest that reflected the persistence of anxiety about military power. Representative Dutee J. Pearce of Rhode Island summed up legislators' general suspicion of those in the military establishment when he cautioned that "even in these piping times of peace, it behooves us to have a care lest the military steal a march of us." Richard Coke Jr. of Virginia focused more specifically on the dangers posed by a federal government that controlled weapon production. He argued that men fought best with the weapons to which they were most accustomed and opposed any plan to "supersede the use of squirrel guns and fowling pieces generally in the hands of yeomanry of the country, by arms of uniform caliber." His support of familiarity over standardization was challenged by Wiley Thompson of

Georgia, one of the most vocal proponents of the plan to reestablish the independence of the Ordnance Department. Thompson argued that even "unparalleled gallantry would become an easy prey to a well equipped and well disciplined foe," if using weapons of "a variety of calibres."[140]

The Seminoles continued to be one such foe.[141] The same year that Ordnance regained its autonomy, 1832, the United States signed a grossly unequal treaty that required Seminoles to move west of the Mississippi over the ensuing three years.[142] They did not want to go, a fact made brutally apparent by the murder of the officer appointed to superintend their removal several days after Christmas in 1835. On that night, General Wiley Thompson and another officer strolled outside the garrison's perimeters, where a party of Indians ambushed them. Commanding officers stationed at the fort heard the altercation and called out several dragoons, but it was too late. Thompson and the other officer died on the spot. Thompson was shot fourteen times and stabbed in the chest.[143]

It is very likely that taxpayer money, doled out under the tenures of Wadsworth and Bomford, paid for the weapons used to kill Thompson. Archaeological evidence shows that the majority of firearms in Florida in the 1830s and 1840s were manufactured either at a national armory or at the factory of a federal contractor.[144] Ironically, eight months before his death Thompson had tried to prohibit the sale of arms to Indians, only to have his regulation repealed one month later. President Andrew Jackson believed that outlawing gun sales to Indians was futile because they would find guns irrespective of the law. If anything, he argued, regulation would only increase hostility.[145] Whether Jackson or Thompson was right, guns were a problem in Florida.[146] Despite the seeming ubiquity of firearms among Florida Indians, the US government had to make sure its own troops and settlers were adequately armed.

The conflict was difficult to assess, and Secretary of War Lewis Cass told the Committee of Ways and Means that "the means of making anything like a detailed estimate of the expenses, are not within the reach of the Department."[147] Each year, the War Department distributed arms, usually muskets, to state militia in proportion to the number of men in service. In the middle of the war, Florida had about thirteen hundred of its own militia, plus militia from Washington, DC, Pennsylvania, Louisiana, New York, Alabama, Tennessee, Georgia, Missouri, South Carolina, and "friendly Indians." These militia ranged anywhere from a handful to twenty-two hundred (South Carolina and Tennessee supplied the most men), and between two thousand and four thousand federal troops.[148] Supplies were often overdrawn.[149] In many ways, numbers were meaningless. They were difficult to determine because of the nature of warfare in Florida. In 1837, former president Andrew Jackson wrote to the War Department that "a well-chosen brigade with such officers as I could select, numbering 1000 bayonets and rifles, in addition to the regulars now in Florida would destroy the Seminole Indians in 30 days from the time of their reaching Tampa Bay."[150] He was wrong. Indians knew the land in a way that intruders did not. Former Adjutant General and Florida politician James Gadsden complained about their hiding places, and based on the logic of American superiority, he assumed that "when the tomahawk was raised the seminal [*sic*] should have been crushed in six weeks."[151] Instead, the war was "shamefully prolonged" and US failure was a constant source of humiliation for army officers.[152] Gadsden was "embarrassed, as an American, at how ineffective we have been against them . . . [that] a small band of savages should have been able to avoid every force sent against them."[153]

Adaptation was the only way to deal with this embarrassment. The Ordnance Department learned to supply weapons for guerrilla warfare.[154] Officials employed, for example, buck and ball cartridges, because they dispersed more widely than traditional ones and were best for camouflaged fighting among swampy forests.[155] For this reason, the War Department placed increased importance on the improvement and experimentation of arms during the late 1830s. In 1837, a board of officers conducted a series of experiments on guns produced at private and public armories. Their comments on the trials reveal a preoccupation with combat in Florida. The board noted that one particular rifle was superior because it could transition between infantry and cavalry use seamlessly, a capability that was important in Florida, where many operations relied on dragoons (soldiers who fought as cavalry when mounted, as infantry when dismounted). The board was particularly sensitive to on-the-ground conditions; one note specifically mentioned the heat and damp of Florida.[156]

Despite the best efforts of the War Department to improve industrial production and army management, there was no real victory in Florida. Neither was there a defeat—just a winding down of military operations. Instead of negotiating a peace, the commander of US troops offered remaining Seminoles money and a rifle to move to a reservation in Southwest Florida, while Congress passed a law in August 1842 for the "armed occupation" of Florida by settlers who would receive federal money to defend their land. Most settlers, however, wanted statehood because they recognized that the national state was the best guarantor of weapon security and military protection against the diminished but still effective Seminoles.[157] They got their wish when Florida became a state in 1845. Persistent violence erupted into the Third Seminole War a decade later, but by the eve of the Civil War four thousand Seminoles had been forced to remove to Oklahoma. Only three hundred remained in Southwest Florida. Although never defeated on the battlefield, Seminoles lost out to white settlement and Florida took its place among the mythical "Old South."[158]

The Future of Stimulus

Not all forms of federal stimulus had similar outcomes. Diplomacy with the disintegrating Spanish empire resulted in a windfall for specific merchant-industrialists that vastly exceeded what tariff protection offered other manufacturers.[159] Similarly, the reorganization of the War Department and its lobbying of Congress in the wake of the War of 1812 cultivated a select network of arms makers. After the United States waged two costly and expensive wars against the Seminoles, it would go to war again in the 1840s in Mexico—its first "official" war since the War of 1812.[160] The governor of Florida continued to solicit weapons and men for the "speedy removal of the remnant of the tribe of Seminole Indians yet remaining in Florida."[161] Recurrent warfare depended on the nation's ability to equip its soldiers readily; industry had to keep pace with expansion and war.

The ability to supply the nation with the stuff of war became a reality because of federal stimulus spending. The post–1812 propulsion of the American arms industry to the forefront of the international market derived from federal responses to territorial expansion and warfare in the Southeast. By the 1820s the New England arms industry, under the oversight of increasingly assertive ordnance officers, achieved unprecedented levels of efficiency and quality, which US presidents complimented enthusiastically.[162] Congress's ability to eradicate the Ordnance Department's sovereignty proved that the government's system of checks and balances worked, but in many ways this system was more illusion than reality. Ordnance officers overcame legislative barriers to what they deemed sufficient production levels by convincing Congress to increase the budget for arms manufacture or, more often, by circumventing limited appropriations with clever accounting or blatant overspending. The financial support that the Ordnance Department provided small arms contractors made possible the manufacture, improvement, and diversification of weaponry to fit national demands.[163]

The textile industry, too, had improved in the years since Tench Coxe, as purveyor of public supplies, implored American manufacturers to improve the quality of their cottons in order to decrease military-related importations.[164] In 1820, Callender Irvine, the new purveyor, still doubted whether American manufacturers could produce the quality or the quantity of cloth needed to clothe the militia so that they were "on equal footing" with the regular army.[165] Yet before long, that quality and quantity would indeed increase. The Merrimack Valley, flush with fresh capital, saw the proliferation of large integrated factories that each specialized in a different kind of cloth. These new factories churned out browned and bleached shirtings and sheetings, jeans, flannels, ginghams, and calicoes that could outperform imports from Britain, Russia, and Germany in domestic markets. Their wares offered ordinary consumers and military supply contractors alike an array of good quality and inexpensive choices. The Boston Associates' good fortune in Spanish Florida had helped make available the made-in-Massachusetts calicoes that westerners could now buy or the flannels that soldiers could now wear.

Florida, and expansion more generally, provided the capital and the impetus for growth for northern manufacturing. Under the patronage of the War and State Departments, the newly expanded and improved textile and small arms industries could now supply goods for the nation. Home markets, though, would not and could not be enough as factory growth led to increased supply and lower prices. New markets had to be sought elsewhere.

South America at the Start of the Independence Wars. From Louis Stanislas D'Arcy Delarochette, *Columbia Prima or South America* (London: William Faden, 1811).
Courtesy, American Antiquarian Society

[5]

Managing New Markets

WILLIAM TUDOR WONDERED whether an exclusionary duty on cotton cloth made in the United States, fast becoming a common item in Latin American markets, was "an issue to be negotiated, or something that should be left to the manufacturers to overcome."[1] In the spring of 1825 Tudor, the US consul to Lima, Peru, chose the former. He alerted Secretary of State Henry Clay that this issue "must be a decision for the government."[2] Rather than subject American textiles to market competition, Tudor met multiple times with the council of government in Lima and brokered a more favorable tariff policy for American exports.[3] Tudor had proven himself right: "Whenever the United States have a representative here with diplomatic powers, [it can] prevent insidious regulations."[4]

The development of early republican capitalism required a body of public servants like William Tudor to promote commercial interests abroad. Throughout the 1810s and 1820s, US federal agents gradually transitioned from a policy of relationship building and information gathering to one in which they solicited new governments for regulations favorable for US manufactures. Tudor and other agents pursued a trade relationship in which the United States exported its manufactured wares in exchange for the raw materials that it could not produce itself. The process by which this exchange became a reality cannot be explained by free trade, economic rationality, or comparative advantage, but by US diplomatic policy in Latin America.[5] While commercial diplomacy mattered everywhere, it is to Latin America, during its independence era specifically, that we must look in order to understand the rise of US industrial power.[6] There, consuls like Tudor navigated the United States' transition from merchant to industrial capitalism. As US diplomats worked on behalf of public and private manufacturers, they took advantage of Latin American nations' requests for provisions and diplomatic recognition in their pursuit of independence from Spain.[7] Although Latin American governments needed US arms in the 1810s to wage war against colonial rule, they did not need American-made cassimeres and broadcloths in the 1820s.[8] By then, however, negotiations pertaining to arms exportation and political support in the 1810s had set a precedent for US advantage in Latin American markets.

Rivaled only by Great Britain, the United States achieved a commercial and diplomatic dominance in Latin America first in relation to arms.[9] In the early 1800s, most of the world's firearms came from Western Europe; the Napoleonic Wars, however, created a shortage in international supply, and Spanish American patriots turned to the United States, which by 1810 manufactured most of its own arms and produced a surplus of gun parts every year.[10] Although repeated pleas for loans and military reinforcement from the United States often fell on deaf ears, wartime demand for guns shifted the policies of Latin American governments in favor of US trade, paving the way for commercial diplomacy that benefited US textile sales, for which no obvious demand existed. The power dynamics that emerged as a result of military provisioning subsequently enabled US agents to push aggressively for favorable textile regulations in the mid-1820s. These dynamics can be understood as the beginning of the United States' exertion of soft power in the region, whereby the United States used its growing economic power to pursue ever more favorable trade policies.[11] International markets had been an important part of the American economy since before the Revolution, but access to South American markets became increasingly important for US manufacturers, whose exports almost exclusively went to the Caribbean islands, Mexico, and South America.[12] When the State Department surveyed American manufacturers in the early 1830s, 105 factories—or about 40 percent of the factories capitalized at more than $50,000—declared that they exported a significant portion of their produce to South America.[13] Diplomatic power made possible this exportation.

After most Latin American republics achieved independence, the United States formally recognized their sovereignty by assigning ministers, who, unlike consuls, had the power to enter formal treaties. These ministers continued the diplomatic work that consuls had done in the 1810s and early 1820s, when the region, once closed off to most US exports, became a profitable outlet for American manufactures. From 1816 to 1822, US exports of domestic produce to Latin America rose from $3 million to $5.3 million, and again to $6.2 million in 1836; in the 1830s, the region received one-third of all the United States' exports, the majority of which were manufactured goods.[14] By the 1840s the United States had entered into several favorable treaties of amity and commerce with South American nations and regularly sent cargoes of manufactured goods in return for minerals, hides, dyewoods, and other goods necessary for the United States' transformation into an industrial power.[15]

New Opportunities?

By the time the Napoleonic Wars reached South America, the United States had done much to remedy what Alexander Hamilton once referred to as its status as "victim of the system." Treaties with European nations and manufacturing initiatives at home brought the United States closer to commercial parity with Europe. But while the United States was no longer "precluded from foreign commerce," its goal of access to ample markets in which to dispose of American produce remained elusive.[16] South American independence offered the United States the opportunity to make its goal a reality. During the Independence Wars that racked Spanish and Portuguese America from 1808 through the 1820s, US diplomatic agents ostensibly maintained neutrality, at the same time as they took advantage of Latin American requests for material and political support in order to achieve favorable market conditions for US goods.[17]

The termination of colonial relationships put the United States in a positive position both commercially and diplomatically. Spain had long maintained royal monopolies by granting favorable commercial contracts to Spanish merchant houses and levying onerous duties and restrictions on foreign vessels. Early letters from the US consul in Havana complained of the "jealousies that appear by the officers of government against our trade" and the expectations that US trade with the island would "dwindle away to nothing."[18] This would change in Cuba, as Spain came to rely on the United States to provision an island dedicated almost solely to sugar production. Before the Independence Wars broke out, though, this reliance did not extend to other colonies, or to trade in anything other than lumber and foodstuffs.[19] For all other commerce, the Spanish Crown believed in a "free," government-protected trade between European and American Spaniards, which operated to the detriment of American traders seeking outlets for European and Asian goods, as well as for North American produce like tobacco.[20] Monopolies were especially burdensome for merchants seeking silver, one of the most valuable commodities of the early modern world.[21] By 1800, Spanish silver constituted 80 percent of legal tender in the United States, but merchants could not legally acquire it directly from South American mines because of imperial trade policy. The only sector in which Spain had encouraged truly "free" trade was slave trading. In 1791, Spain began liberalizing its slave-trade policy by increasing tonnage limits and the amount of time a vessel could remain in port, enabling foreigners to supply slaves for Spain's colonial sugar plantations. Once the Napoleonic Wars broke out in 1803, however, British naval power cut off the Spanish dominions from African sources of supply, causing the virtual cessation of traffic in slaves over the next few years. While Americans certainly took advantage of this short-lived opportunity in trafficking human bodies (and would continue to do so illegally), they soon needed to find other commodities to trade in South America, especially since many of Spain's former colonies enacted legislation that restricted slavery and the slave trade.[22] Consuls helped find these new commercial opportunities, which soon included manufactured goods.

The Spanish American Independence Wars made it possible for the United States to move into the diplomatic vacuum created by Spain's diminishing power. During the 1790s and early 1800s, the Spanish Crown denied foreign nations consular presence in any of its colonies, except Cuba and one post at La Guiara in the viceroyalty of New Granada. Portugal, however, allowed the United States to maintain a minister in colonial Brazil at Rio de Janeiro. As soon as Buenos Aires deposed the Spanish viceroy and established an independent junta in May of 1810, the United States appointed Joel Poinsett "special agent of the United States to South America." He was soon followed by commissioned consuls in Buenos Aires, La Guiara, and Valparaiso. These men occupied nebulous positions. In fact, Special Agent to the United Provinces of Rio de la Plata William G. D. Worthington specifically requested that his passport be "general, so I go as not an agent but a citizen of the United States."[23] Because consuls could not enter into treaties with foreign nations (as ministers could), they were able to assure South Americans of their "real," albeit unofficial, affection for the patriot cause. These unofficial assurances set the stage for favorable commercial relations.[24]

The United States had been appointing consular representatives to important port cities since its declaration of independence. The first representatives worked to secure commerce with France during the American Revolution, and by the 1790s the United States had consuls in such far-flung places as Canton, Tenerife, Franconia, Gibraltar, and Cape Town.[25] The ratification of the Constitution created an executive with the power to expand the consular service, and George Washington and Thomas Jefferson were two of the strongest advocates of a foreign service that could promote American trade in a way that congressional legislation could not.[26] In his second annual address to Congress, President Washington declared that "the patronage of our commerce . . . has called for the appointment of consuls in foreign countries."[27] Consuls were appointed by the president at the advice of the State Department, which regularly fielded solicitations from merchants seeking consular posts for themselves or their friends.[28] Candidates for the position had to demonstrate their familiarity with the particulars of their host ports' markets: the goods in demand, the market prices, the fees and tariffs, and the issues with safety. Most important, they were expected to be successful merchants. Application letters attested to the individuals' success and reputation in the commercial world. During their tenure, they were not only allowed, but encouraged, to profit from trade. In 1778 Benjamin Franklin, John Adams, and Arthur Lee had suggested in a letter to Congress that consuls should be "better qualified for this business than any others . . . and the advantages to be derived from trade, will be a sufficient inducement to undertake it."[29]

Consuls' success in the commercial world was inextricable from their service to their country and to their constituents. As US consul to Havana John Morton wrote in 1801, "Having been considerably engaged in business, I have experienced the additional advantage derived from acting in a public capacity, knowing that I had the additional weight or influence with officers of the courts of justice."[30] Historian Ellen Hartigan-O'Connor has shown how the appointment of public auctioneers in the early republic raised fundamental questions about federal patronage and the common good.[31] While the appointment of consuls, unlike that of auctioneers, remained largely outside the purview of ordinary citizens, the same conflict and complementarity between private interests and the regulation of the economy were at play. Consuls were, at the same time, well-connected merchants and agents of the federal government charged with, in the words of an instruction guide, giving their "attention to whatever can promote the commerce and navigation of our country, as well as the particular affairs of the individuals of our nation who may require the exercise of the consular function."[32]

Acting as both state agents and private merchants, consuls worked toward US commercial objectives in South America. These objectives included the minimization of embargoes and blockades, the reduction of tonnage duties, the adjudication of violations of law, the settlement of shipping claims, and the garnering of equal or preferential treatment in tariff legislation. Prior to the establishment of official diplomacy, the United States could not enter into trade treaties, but it could achieve some of its objectives without endorsing independence. Consuls were particularly effective at navigating shipping claims. Merchants sent countless letters to the State Department requesting aid in pursuing their legal claims against various foreign governments. These petitions requested remuneration or exemption from payment in response to a variety of grievances, which included the perceived unlawful seizure of a vessel or cargo or the application of a particular duty or custom house fee, as well as damage to property that resulted from the insufficient policing of waters and territory within a government's jurisdiction. The department delegated much of the work of claims cases to their consular agents abroad because consuls were the bureaucratic gatekeepers to the foreign governments and to the information necessary for indemnity. Consuls' success in achieving favorable settlements allowed the secretary of state to assure prominent merchants that the US government was on the job.[33] Efficacy in this arena of diplomacy meant that some merchants sidestepped the State Department altogether, saving the department unnecessary paperwork. New York shipping firm Baldwin and Spooner, for example, wrote directly to US consul John MacPherson, who also acted as their commercial contact in Cartagena, where he served as their agent in pursuing claims against a Colombian governor for seizing a quantity of tobacco that had already passed lawfully through customs at Chagres.[34]

Consuls wielded American commercial power, which was in many ways more alluring than the military might of imperial European nations when dealing with newly independent nations in the Western Hemisphere. As Supreme Director of Rio de La Plata Gervasio Antonio de Posadas explained in a letter to President James Madison: "The greatness of the powers of Europe has been founded on our degradation. It is in you we place our present hopes." In the wake of European mercantilist control of Latin American trade, officials recognized the importance of US markets and shipping for economic independence. Posadas wrote to Madison informing him that US "influence in the commercial world" prompted his government to seek US support.[35] These requests helped the early republican United States achieve access to new markets, while also beginning to exert its commercial power in the region. Firearms were a major source of this power.

Selling Arms in an Age of Revolutions

At the same time as the United States fought its second war of independence against Great Britain, South Americans fought their first against Spain and Portugal. But while the United States had established a domestic arms industry, its neighbors to the south had not. Patriots and royalists alike, in fact, turned to the United States for weapons in the face of European shortages. Throughout the 1810s, the United States' ability to maintain, expand, and improve its own arms supply increased, as did its production of superfluous and substandard guns. Diplomatic considerations prevented their export from being a foregone conclusion, but once federal officials and individual merchants created an arrangement whereby private parties were able to sell public arms, US manufacturers disposed of surplus stock in Latin America. As diplomats began to cultivate a sense of material and political dependency among patriot leaders in the 1810s, they laid the foundation for solid textile sales in the 1820s.

The Independence Wars appeared especially advantageous for US arms makers and sellers, but they at first presented a conundrum for commercial policy: To whom, if anyone, should the United States provide weapons of war? Soon after Venezuelan patriots established an independent junta in 1810, the Spanish Captaincy General of Venezuela gave a New York vessel the exclusive privilege of loading and shipping her cargo, which included weapons for the royal army.[36] Likewise, the patriot government at Buenos Aires repeatedly sought arms from the United States, asking for "10,000 at a fixed price or the permission to export from the United States that number of arms."[37]

Supply, though, did not always meet demand. First, many Americans did not want to sell to Spain. They opposed the provisioning of a European empire that they deplored as oppressive and tyrannical, and instead demanded that the United States recognize and support the patriot cause.[38] US merchants in Colombia specifically requested that the United States prohibit the exportation of arms to Maracaibo, where the Spanish army stockpiled military supplies.[39] Second, although no international law prevented the United States from providing guns to rebel governments, the official military provisioning of Spain's colonies would hinder diplomacy with Spain, derailing negotiations for the acquisition of Florida.[40] Civilian merchants, though, offered a way around this challenge, a solution Thomas Jefferson had anticipated in 1793 when he declared that private persons be permitted to export arms to belligerents.[41] Jefferson's pronouncement became US policy for most of the nineteenth century, and New England arms manufactured under government patronage, either by the federal armories or by private contractors, made their way to Mexico and South America. By allowing individual merchants to sell both public and private arms to whomever they chose, the US government made it possible for Latin American patriots to wage war, and for US diplomats to obtain favorable trade policies as a token of their indebtedness, all while appeasing the Spanish government.

The arrangement by which private parties exported government arms, while US agents on the ground maintained the official policy of neutrality in response to requests for aid, worked out well for both the United States' international reputation and American manufactures. It is difficult to know just how many arms went to Latin America and by whom they were produced, because annual lists of exports for the United States for the 1810s and 1820s do not specifically list types of weaponry, even for years when we know US arms were sold in South America. The international arms trade is almost always shrouded in controversy and secrecy; the US trade during the Latin American Independence Wars was no exception.[42] But many of the US arms that were exported likely came from a New England manufactory that depended on government patronage for its existence. As we have seen, the small arms industry was located mainly in this region and would not have existed but for the government contract system.

Yet while government demand afforded arms makers a more reliable market than private consumers could provide, manufacturers occasionally had to look elsewhere for sales. For example, Nathan Starr, the owner of one of the nation's oldest private arms manufactory in Middletown, Connecticut, and contractor to Springfield Armory, exported arms to Colombia following the War of 1812 to supplement his income.[43] Starr, whose factory operations depended on federal contracts, claimed that the exportation of surplus arms compensated for the depreciated wartime payments he had received from the US government.[44] And despite guaranteed domestic demand and financial support, the national armory at Springfield also produced a surplus of arms each year. Most of this surplus was disposed of through agents and auction houses in Philadelphia and New York, whence it was sold to domestic patrons or to Latin American patriots, often with the knowledge of federal officials.[45]

In addition to providing the capital for the arms industry, the federal government nurtured commercial relationships with Latin Americans that facilitated the ease with which these arms were sold. In response to patriot leaders' requests for US arms and political approval, US State Department agents cultivated good will by assuring them of US sympathy for their cause, which helped create a commercial environment in which shipping claims were settled and preferential treatment in tariff legislation was secured. US diplomatic agents understood well that the United States' status as a first-rate producer of armaments gave it the upper hand in negotiations. A representative from the newly established independent junta at Caracas declared to US Secretary of State Robert Smith that New Granada's harbors would "await [US trade] with open arms."[46] Foreign markets were indispensable for a political economy geared toward disposing agricultural, and increasingly, manufactured, produce. Supplying arms to revolutionaries not only refilled federal coffers; more broadly, it served national economic interests. As one consul reminded Secretary of State James Madison, "Should this country succeed, their trade with the United States will be lucrative and important, but should it for the want of assistance succumb, the ports will be shut against foreign trade."[47] Arms were the first step toward making sure the ports did not shut.

Examining the relationship between arms exportation and diplomacy in one recipient country allows us to understand how unequal power dynamics emerged around demand and supply, making possible the type of diplomatic relationship that provided a boon to American manufactures. The independent junta in the viceroyalty Rio de la Plata (present-day Argentina, Bolivia, Uruguay, and Paraguay) was particularly solicitous in its arms requests. In 1812 one consul wrote to Secretary of State James Monroe that the government wanted to ship a large amount of copper to Boston in exchange for arms; another that patriot leaders wished to purchase ten thousand stands of arms.[48] Military deficiencies set the stage for an unequal trade relationship in which Supreme Director of the Provinces of Rio de la Plata wrote to President James Madison that if the United States would supply arms, his government would "be ready to engage in any treaties of commerce that will be advantageous to the United States" and "duties on US manufactures would be reduced."[49] Former depots of Iberian produce now opened to US wares.[50]

The patriot forces in Rio de la Plata were especially receptive to Springfield arms. The superintendent of the Springfield Armory maintained a complex web of dealers in North and South American cities to facilitate the sale of arms that domestic markets did not absorb. When the armory exceeded its annual budget, which it often did, the War Department required it to dispose of excess materials to make up for overproduction.[51] While the armory often sent surplus stock to New York or Philadelphia, not all of its articles always sold. The superintendent, in fact, looked for additional outlets in the new state of Mississippi, but found that rifles only sold at low prices and that a "more advantageous sale could not be forced."[52] Additionally, the government did not want any more paper money from "south of Philadelphia"; the specie that selling agents could obtain from South America was a more satisfactory mode of payment.[53] Even in an expanding United States, there was not enough demand for firearms.

In order to transition into foreign markets, Springfield superintendent Roswell Lee cultivated a mutually beneficial relationship between the federal government and selling agents: the agents found outlets for surplus stock and gave the armory direct feedback on their products, while the agents profited from US diplomatic presence abroad. One of the Springfield Armory's principal agents in Philadelphia was William Cramond, a merchant who recognized that surplus arms "were of little use to the United States" and that a "low price for them might induce a sale of the whole."[54] Lee perhaps chose Cramond as his agent because of Cramond's experience with South American commerce and his ties to merchants in Buenos Aires.[55] Cramond was connected with Baltimore shipping firm D'Arcy and Didier, which dealt government arms both in Argentina and Mexico in 1815 and 1816.[56] In July 1815, Cramond received authorization from Captain John Morton of the Ordnance Department to negotiate a sale of arms with Lee. There was some sense of urgency in this deal. Cramond sent his agent George Hockley to purchase arms with a letter from the acting Secretary of War Alexander Dallas, who had approved the mode of payment so that "no delay in shipping the arms may take place."[57] In a series of visits, Hockley negotiated with Lee for a fair price so that business could proceed.[58]

From his agents, Lee received information about Latin American markets that would enable him to make the best possible decisions about arms exports and thus improve armory sales. He knew, for example, that the sale of additional arms to Argentina would be advantageous. Cramond informed Lee that one agent in Buenos Aires had received favorable reports from purchasers of old Springfield arms in 1816.[59] Lee also knew that shorter-barreled muskets "were in very high demand in South America" and were in fact favored over long-barreled ones. But in order to sell these weapons at a profit, the selling agent needed from Lee specific guarantees that they had been properly inspected by the armory's inspectors and were found to be of good quality, "though probably not of the first quality," and that they were being sold only on account of the size of the barrel. Lee also had to inform the agent of the price difference between the two types, as short-barrel muskets were not usually made for US service.[60] This type of communication enabled Lee and his merchant to profit from muskets that US soldiers could not use. Because selling agents alerted Lee to quality issues and preferences and solicited production information for marketing purposes, patriots in Buenos Aires bought them in "preference to others and at the highest price in the market."[61]

These arms vendors, meanwhile, benefited from the aid of US diplomatic agents in South America. The shipment of arms promoted positive opinions of the United States and created a situation by which the United States achieved commercial power as the main supplier of military goods; exportation alone, however, was not necessarily enough to secure business deals for American manufactures. To be sure, patriot leaders readily accepted and appreciated armaments. The US consul at Buenos Aires in 1812 informed Secretary of State James Monroe that a small supply of arms had arrived on the *Liberty* and increased the sense that the United States should be "looked up to not only by the government but by the people as the only sincere friends."[62] Throughout their struggle against Spain, Argentinian leaders echoed this sentiment in letters to US presidents.

Even as South Americans sought business with US merchants, these merchants still needed federal help. John Devereux, a Baltimore merchant, relied on the US consular agent at Buenos Aires to execute his contract with General Juan Martin de Pueyredón for military supplies and a loan of $2 million.[63] US consul Thomas Lloyd Halsey was an ideal commercial agent because he resided in Buenos Aires and was on good terms with the leaders there, who had given him assurances that US commerce would always benefit from favorable treatment.[64] Likewise, Baltimore shipping firm D'Arcy and Didier, which purchased and sold Springfield arms, maintained a close relationship with the US consul at Valparaiso in order to better facilitate their trade there.[65] This consul, Henry Hill, informed the firm of blockades, duties, and other pertinent commercial information. Hill's correspondence with them and other merchants helped mitigate the transoceanic effects of imperfect information—an invaluable service in the United States' transition to capitalism.[66] When the patriot government in Valparaiso was unable to pay for all the supplies it needed, the government removed tariffs on military items, which encouraged selling houses like D'Arcy and Didier to ship more arms. After D'Arcy and Didier's cargo of three thousand muskets arrived, Hill alerted the Supreme Director, who then issued a public announcement calling on all citizens of the state to purchase these particular arms.[67] Weapons may have been duty-free, but their sales were not guaranteed. What if, for example, a cargo of arms was captured at sea? In 1819, a schooner from Baltimore carrying a cargo of "sundry merchandise" and arms was captured off the Port of Callou by a Chilean squadron commanded by British Admiral Lord Cochrane. US consul Henry Hill immediately petitioned the Tribunal of Prizes in Santiago for the ship's release.[68] Agents like Hill and Halsey helped make South America an ideal place to send surplus armaments.

Post-Independence Markets for Textiles

Consuls' commercial support was essential not just for the sale of arms in Latin America, but also for the diplomatic precedent it set for the exportation of US textiles. As diplomatic agents facilitated sales and assured patriots of US support for independence, they built up good will and helped establish the United States' reputation as an important trading partner, which made it possible for diplomats in the 1820s to push for lower tariffs and favored-nation status. In the absence of official diplomatic relations, Halsey and others made business deals and helped cultivate favorable opinions of the US presence in the region among Latin American officials and businessmen. Halsey, for example, was invited to visit Uruguay's Chief of the Oriental Republic Jose Artigas and was received with "marked attention and kindness both by the chief and by all classes of people." Halsey's boasts do not, of course, necessarily reflect the actual sentiment of those around him, but they do suggest that he was treated with a level of respect that would permit favorable negotiations.[69] Prior to the commencement of official diplomacy, this cultivation of amicable relations both created a stable and friendly commercial environment for American merchants and laid the groundwork for later policy implementation. US diplomacy in Latin America in the 1810s had promoted the American arms trade within the limits of neutrality. In the wake of independence, the diplomatic and commercial climates changed. The ability of New England textiles to compete successfully with British ones required a more aggressive strategy than had been possible or necessary in the 1810s.

After years spent staving off political recognition of patriot governments, the US government finally commenced formal diplomacy with the new nations to its south. In 1822, Washington began receiving representatives from various Latin American nations, and the State Department established official diplomatic missions in return.[70] Finally, according to President James Monroe, it "seemed obvious" that Spain could not subdue her former colonies, nor "improve the condition of the people in those countries," and so official neutrality gave way to a more forceful assertion of US hegemony in the hemisphere—the Monroe Doctrine, through which the president declared the hemisphere closed to "future colonization by any European powers."[71] The US government could now make pretensions to being the dominant "interest" in Latin America—a position that US consul to Buenos Aires William Worthington had urged John Quincy Adams to assert five years earlier. Worthington had warned that if the United States did not take a more proactive stance in its recognition of independence and in its formal commercial policy with the southern provinces of South America, Europe would gain too strong a foothold and the United States would "lose all the glory and profit to be desired from this country."[72] Worthington need not have worried too much. By 1832, Americans had more ships in many South American ports than any other nation and its "tonnage was daily increasing."[73]

Yet the increase in American commerce was not a foregone conclusion, and in the 1820s, there were still barriers to overcome. The military urgency for provisions diminished, and new governments enacted heavy tariffs to recover from wartime debts and encourage domestic agriculture and manufacturing. Although Latin American leaders had specifically solicited American arms, they did not request American textiles. Nor did they allow textiles to enter ports duty-free as they did with military stores. The introduction of American textiles required more diplomatic work than had arms exportation because imported cloth ran up against these barriers to trade. The congress in Vera Cruz, with whom the United States had an extensive overland trade, for example, prohibited in the early 1820s the importation of all cotton goods.[74]

The two main barriers to entry for US textiles were tariffs and subsidies for domestic manufactures in Latin American countries. Following independence, some Latin American governments began experimenting with targeted incentives for domestic manufacturing. The first documented case of government support for industry was an 1825 Mexican act that exempted paper manufacturers from excise taxes, and by the 1830s several countries had adopted similar encouragements.[75] Peru developed an industrial sector in its northern provinces that enjoyed nationalist economic policies in the capital city, where even city consumers were not wholly opposed to cost-raising tariffs on imports.[76]

Latin American scholars have debated the extent to which post-independence Latin American governments adopted protectionist trade policies. Beginning in the 1980s, economic historians began challenging the arguments put forth by dependency theorists, in which liberal elites in Latin American nations imposed detrimental free-trade policies that rendered their countries dependent on Europe. Historian D. C. M. Platt argues that the region was unwillingly self-sufficient in the nineteenth century and that an inability to pay for foreign goods produced strong incentives for home manufacturing. John Coatsworth and Jeffrey Williams, on the other hand, argue that the revenue demands of almost-constant military conflicts led most Latin American governments to levy high tariffs, so that by 1865 the region had some of the highest tariff rates in the world.[77] Other approaches emphasize the internal politics and regional differences that made policy implementation anything but uniform throughout the region.[78] Colombia, Chile, and Argentina had particularly low tariff rates, while Mexico, Peru, and Brazil experienced more protectionist policies, but even these rates are debated among scholars.[79] Protectionism varied depending on local geography, domestic politics, world price rates, and international relations. Regardless of the precise levels of duties, however, Latin America was far from being a region that threw open its doors to liberal trade—a fact of which US consuls were quite cognizant. The US consul to Vera Cruz, William Taylor, made Secretary of State Adams aware of these impediments to American trade, specifically the prohibition of cotton imports, and suggested that a minister with full treaty-making powers stationed in the country would improve the situation.[80]

After the US government heeded Taylor's request by appointing Joel Roberts Poinsett as the nation's first minister plenipotentiary to Mexico in 1825, American cottons found a "ready vent" among Mexican purchasers. But even as Poinsett reported that some consumers were starting to prefer US cloth to British, he urged the improvement of American domestic manufactures.[81] As trade redirected from Spain and Portugal to northern Europe and the United States, Latin American consumers still desired the latest fashions, which did not exactly bode well for US manufacturers whose own domestic customer base still often preferred European wares to American ones.[82] On the other hand, Latin American nations had a substantial body of poor consumers whose needs were not met by fine European imports or upstart local firms catering to elite preferences. This created a space for US goods, especially as many factories specialized in coarser fabrics, like cheap broadcloths and kerseys.[83]

US consuls helped manufacturers at home navigate these preferences, just as the output of textile factories was increasing. The timing of Latin American independence and the commencement of formal diplomacy boded well for the expansion of manufacturing. At the same time as the United States began acknowledging national sovereignties, for example, the stockholders of the Merrimack Manufacturing Company in Lowell, Massachusetts, the largest manufacturing undertaking of its time in the United States, voted to improve water power and expand production.[84] They found new markets in South America, in part because they had a consul working for them.[85] The case study of one consul's negotiations with Peru, a nation with its own burgeoning industry and protective economic policies, illustrates the degree to which diplomacy mattered for US exports and the rise of industry. If we are to understand the early success of the Waltham-Lowell System we must do so in the context of diplomacy. US consul to Lima William Tudor successfully lobbied the government of Peru for measures that specifically benefited his friends in manufacturing. While exports to Latin America accounted for just one part of their profits in the 1820s, the ability to export mattered to them and so it matters to the story of manufacturing in the United States.

The Boston Manufacturing Company in Waltham, Massachusetts, had been in operation for a decade when Nathan Appleton, one of the company's proprietors, sent cloth samples to William Tudor, a leading Bostonian who had been appointed consul to Lima in 1823. Tudor promised to "do his best to introduce them in Lima and Upper Peru."[86] Tudor's seven-year tenure in Peru served both diplomacy and the economy. A royalist stronghold, Peru was one of the last Latin American countries to receive US recognition: Tudor was the only US diplomat there until the government appointed a minister plenipotentiary in 1826. He cultivated amiable relations with his host country and functioned as a liaison for American commerce in Peru. He also worked as an agent for his associates in manufacturing. While Tudor advocated on behalf of American commerce generally, he was most effective in influencing policies that benefited US manufacturing, and specifically those that benefited his friends. Tudor's tenure as consul was in many ways emblematic of early national state-promoted capitalism: his actions stimulated US economic growth generally, but privileged certain individuals.

The men responsible for the Waltham-Lowell System had access to the state's diplomatic network in ways that the owners of most New

England factories did not. The Northbridge Cotton Manufacturing Company in Northbridge, Massachusetts, for example, was average-sized—capitalized at $150,000—and employed the "putting out" system, whereby factory work was outsourced to households or small workshops. It sold locally, and like many others, did not have overseas accounts or correspondence with agents of the state.[87] While many of these companies did just fine without those international connections, they did not become emblems of the new American factory system. The ability of American textile manufacturing to reach its large-scale potential depended on diplomacy to create outlets for their goods, and the proprietors of the various Waltham-Lowell firms had decades of experience dealing with US commercial agents abroad. Israel Thorndike, one of the directors of the Boston Manufacturing Company, had a long-standing account with consul Thomas Lloyd Halsey, who also facilitated arms deals. Halsey had worked on Thorndike's behalf, managing his sales of ox and horse hides and foreign satins, in the years that Waltham goods were still being sold in domestic markets.[88] The Boston Manufacturing Company also had accounts with John Murray Forbes, who was stationed at Buenos Aires from 1820 to 1831 in various diplomatic posts, and was appointed to the high post of chargé d'affaires in 1825. Forbes settled claims related to military stores, achieved reduced tariffs, and maintained friendly relations with both Argentinian officials and the business community. These services were invaluable to American interests abroad. Tudor, however, did more than most consuls to aid the development of American textile manufacturing.[89] Throughout his tenure as consul, he not only served as a sales agent for manufacturers and achieved changes in tariff policies. After the United States recognized Peruvian independence by appointing a chargé d'affaires in May of 1826, he also began to pressure Peruvian officials to abandon domestic manufacturing projects in favor of importing US textiles. Tudor's promotion of US manufacturing helped complete the United States' transition from neutrality to a more aggressive commercial diplomacy.

Tudor spent the first part of his tenure continuing the practice of neutral diplomacy, while also dispensing commercial information and facilitating business deals in Peru. By the spring of 1824, when Tudor showcased Waltham goods among Peruvian shopkeepers, the United States had officially recognized Colombia, Chile, Mexico, and Brazil, but Spain's military success in Lima required Tudor to uphold pretensions to nonalignment. Even as he assured the Spanish viceroy of Lima that the United States had never recognized the Republic of Peru and that his job was strictly to assist destitute seamen and promote the commercial interests of his country, the Spanish viceroy refused to recognize his status as consul.[90] This diplomatic inconvenience, though, did not prevent Tudor from acting as business agent for Boston capitalists. Just as commercial agents had informed Superintendent Roswell Lee about the most suitable arms for Argentine markets, and just as US consul Henry Hill had helped sell Springfield arms in the 1810s, Tudor helped market various styles of Massachusetts textiles. The information he relayed to manufacturers enabled them to make calculated production decisions. Tudor instructed Appleton that coarse cloths would do well among the poorer classes, while "quality" pieces would receive the "same preference they have obtained at home."[91] This information was especially useful to Appleton, who had determined that each of the Boston Associates' industrial undertakings would specialize in different kinds of cloth. The Merrimack Manufacturing Company, which had only just begun operating when Appleton sent samples to Peru, would specialize in fancy calicoes, as would the Cocheco Manufacturing Company beginning in 1827; the Boston Manufacturing Company would always have on hand at least $100,000 worth of coarse cloth.[92] In 1825, construction began on the Hamilton Manufacturing Company in Lowell, Massachusetts. It would specialize in yet other kinds of cloth, which would find ready markets in Buenos Aires and Rio de Janeiro.[93] Knowledge of market preferences enabled Appleton and his associates to be more strategic about production and exportation.

Tudor did more than supply marketing services; he actually improved the market for certain types of cloth in South America. Tudor's initial report to Appleton had informed the manufacturer that bleached cloths would not "answer well" because they were too similar to those made in Peru. Peruvians had begun to manufacture tucuyos, a type of woven plain cotton that often appeared similar to plain white cloth from the United States. Because tucuyos faced competition from cloths imported from England, India, and especially the United States, the Peruvian government levied a prohibitive duty of 80 percent on white sheetings produced in the US.[94] This trade barrier, though, could be overcome with the proper diplomatic maneuverings. In a lengthy letter to Minister of Foreign Affairs José Manuel de Pardo, Tudor argued that the encouragement of tucuyos would drive American domestics out of the market, much to the detriment of the Peruvian consumer.

Despite cosmetic similarities, Tudor insisted on the superiority of American cloths. He maintained that Peruvian manufacturers would not be able to produce the type or the quantity of cloth that the population demanded. Peruvian consumers, he suggested, preferred American cloths to those from India or Britain, which would now be able to dominate the market because of the discriminatory tariff policy. American manufacturers, on the other hand, would adapt their goods to the wants of the country, and if the policy changes happened soon, none of these commercial injuries would come to fruition. In case the minister was not convinced, Tudor also made sure to remind him that half of all the foreign duties that Peru's treasury received came from American trade, lest the minister forget Peru's most important trading partner.[95] Tudor's polite threats worked. The government complied by equalizing the tariff on all foreign goods, a policy change that would enable US cotton goods, in Tudor's estimation, to dominate the market because of their superior quality.[96] The trade in American bleached cloths was not interrupted, and tucuyos, which had had the upper hand in Peruvian markets since at least the early 1820s, lost out to American imports.[97]

Tudor's strategy may have promoted national economic goals of increased market access, but his negotiations centered primarily on commercial policies that would aid his cohorts in Boston. For example, in his negotiations with the Peruvian minister of foreign affairs regarding tucuyos, Tudor made sure to specify that the types of cloth he wanted the tariff change to affect were those made on power looms, the specialty of the Boston Associates (only eleven other Massachusetts factories used them in 1820). Power looms were expensive to procure, and the Boston Associates controlled the patent rights for the best model. Also, irrespective of their price, power looms were not adopted uniformly. Samuel Slater, for example, was slow to introduce the use of the power loom in his mills, not for financial reasons, but because his business model was less growth-oriented than his colleagues' in Waltham and Lowell.[98] As an ancillary consequence of Tudor's targeted commercial diplomacy, however, other manufacturers whose products were similar to those produced by the Waltham-Lowell System also were able to profit.[99] His negotiations ended up mattering for a wider class of manufacturers than just the Boston elite.

Although Tudor was perhaps unique in his pursuit of regulations that supported US textiles—he was certainly unique in his verbosity in economic affairs—other diplomats, too, recognized the importance of trade regulation for textiles. For example, US consul to the Brazilian state of Pernambuco John T. Mansfield observed that coarse cotton goods manufactured in the United States were becoming markedly more popular in the city than English ones and that if assigned the same tariffs as English wares, American imports would greatly increase.[100] Elsewhere, effective commercial diplomacy helped contribute to the ease with which US manufacturers found export markets for their textiles. In his letter to a Boston selling house about the steady importation of domestic cottons into Valparaiso, merchant Charles Frederick Bradford casually remarked that in Valparaiso, "port charges are reasonable; the tonnage duty is only 12.5 cents."[101] This 50 percent reduction in tonnage duties had been achieved by the US consul there several years earlier and would save "many thousands of dollars to [US] commercial interest" over the next decade."[102] And the directors of the Newmarket Manufacturing Company in New Hampshire were pleased to learn that "South America is favorable for domestics and the duties are put on a more favorable footing."[103]

Policy changes such as these were one reason American textiles, and manufactures of all kinds, were fast becoming common items in Latin American ports.[104] And in general, the privileged status the United States and its diplomats had attained as a result of the arms trade and its diplomacy during the Independence Wars created a favorable climate for US exports. Argentina's Minister to Foreign States Bernardino Rivadavia assured US consul to Buenos Aires John Murray Forbes of his good feelings toward him, noting that US recognition of independence mattered more to his government than that of any other nation and that no exclusive privileges would be granted to any nation other than the United States.[105]

A New Relationship?

Although this privileged status abetted an increase in the exportation of textiles from the United States in the 1820s, arms exportation actually decreased as Latin American republics secured independence and no longer required a steady supply of armaments. The arms trade and the federal support it received, however, by no means ceased. For example, New York merchant David Curtis De Forest, whose business connections with the Springfield Armory's agent, William Cramond, linked him with federal supply, sent arms to Peru through the Providence mercantile house Brown and Ives in 1821.[106] And Americans who had exported arms in the 1810s continued to receive federal assistance. For example, merchant Jacob Idler, who in 1817 had contracted with the government at Caracas to supply Venezuela in its fight against Spain, experienced a delay in payment that extended well beyond his contract's stipulations. While this issue was not resolved until the 1830s, the US consul eventually achieved a satisfactory settlement of the case.[107] Similarly, many Americans who had shipped arms to Mexico during the 1810s experienced delays in payment or damage to cargoes, but they were rewarded when Congress established a commission in 1839 to resolve their claims against the Mexican government.[108] The resolution of commercial claims sent the message to manufacturers and merchants, and to foreign governments, that US business interests were protected and promoted, even if they impinged on other nations' jurisprudence.

American claims against the Mexican government, in fact, contributed to the growing hostility between the two governments, as Mexico became a site in which the commercial and diplomatic relationship that had emerged around arms exportation crystallized and became particularly unequal. Throughout the 1810s and 1820s, Mexico relied on the importation of arms from the United States to supply its army.[109] Yet while trade between the two countries became especially important for the economy of the United States, US diplomats, emboldened by their increasing diplomatic power, began more aggressively to assert a policy of commercial dominance.[110] For example, when the cargo of three American brigs was confiscated by a national vessel of Mexico because of the prohibition of imports, US consul William Taylor, who knew that Mexican law prohibited the cargo, called in US military reinforcement to "demand the immediate restoration of these vessels and their cargoes." He justified military action by stating that the United States must command respect.[111] Taylor, who also vouched for the legitimacy of his friends' claims against the Mexican government for the detention of their vessel, despite knowing them to be false, could not countenance Mexican assertion of legal sovereignty if it disadvantaged Americans.[112] Negotiations such as Taylor's increased both the success of American commerce and the hostility with which the Mexican government viewed American commerce.

The assumption of American commercial dominion developed in tandem with greater resource extraction in the region.[113] After denigrating Mexican governance, US consul to Chihuahua John Ward speculated that perhaps one reason for the developing Mexican hostility toward Americans came from "a reluctance to have so much of their monied capital [specie] exported."[114] Although Americans had been importing specie, hides, and dyestuffs from Latin America for decades, these importations intensified in the late 1820s. Logwood, for example, increased as the number of textile manufacturing companies that depended on its bark for purplish-red dye rose.[115] The speed with which raw goods flowed out of the mines and soils of Latin America increased, just as greater numbers of American manufactured goods flowed in.[116] The United States had finally become a threat to British superiority in international commerce. In a show of jealousy, the British minister to Mexico "excluded [US diplomats] from all their social parties and encouraged and gave toasts hostile to the United States."[117] No longer Britain's colonial backwater, the United States developed its own unequal relationship with its southern neighbors.

Soft Power?

This new relationship with Iberia's former colonies marked the beginning of the United States' exertion of soft power in the region.[118] While the twentieth century would see the rise of "dollar diplomacy," whereby the United States doled out loans to Latin American national governments with the implied caveat that these governments adopt US-style democracies and free-trade policies, the 1810s and 1820s were years in which US diplomats and merchants capitalized on the United States' status as a non-imperial nation with growing political and economic power to promote manufacturing interests overseas.[119] By so doing, they set a precedent for the utilization of soft power to achieve industrial superiority. The neutrality, good will, and material disparity that characterized commercial and diplomatic exchange during the Independence Wars gave way to one in which US agents were able to shape commercial policies that eased American manufactures into foreign markets and created demand for American goods.

US firearms had created inequities in the commercial relationship between the United States and its southern neighbors. As South American patriots fought for their independence using US muskets, the Springfield stamps that were proof of standardized quality in the United States became symbols of its manufacturing prowess abroad. Consular agents facilitated the sales of these weapons, which achieved US national security aims doubly: reliable surplus markets made high rates of productive capacity worthwhile for the arms industry, while the development of commercial diplomacy strengthened US political power overseas. For the young US republic accustomed to European dominance, Latin America became one region where it could start to negotiate with the upper hand. Consuls like William Tudor became effective at this sort of negotiation as they brokered policies that would improve the sale of manufactured goods, particularly textiles, in new markets. Private manufacturers who had access to federal agents fared especially well. The policy of economic coercion that diplomats established in Latin America in the 1820s became increasingly important as new textile factories and arms manufactories sprang up throughout New England in the 1830s.[120]

It was in Spain's and Portugal's former colonies that the United States' transition from merchant to industrial capitalism became complete. US agents successfully negotiated the transition from colonialism to independence in Latin America, and the United States emerged as a major supplier of the Western Hemisphere. By the 1830s, as Americans continued to spread out across the North American continent, the United States could arm and clothe not only its own citizens, but those of the new republics to its south. The Western Hemisphere was only the beginning of an industrial manifest destiny that would eventually span the globe.

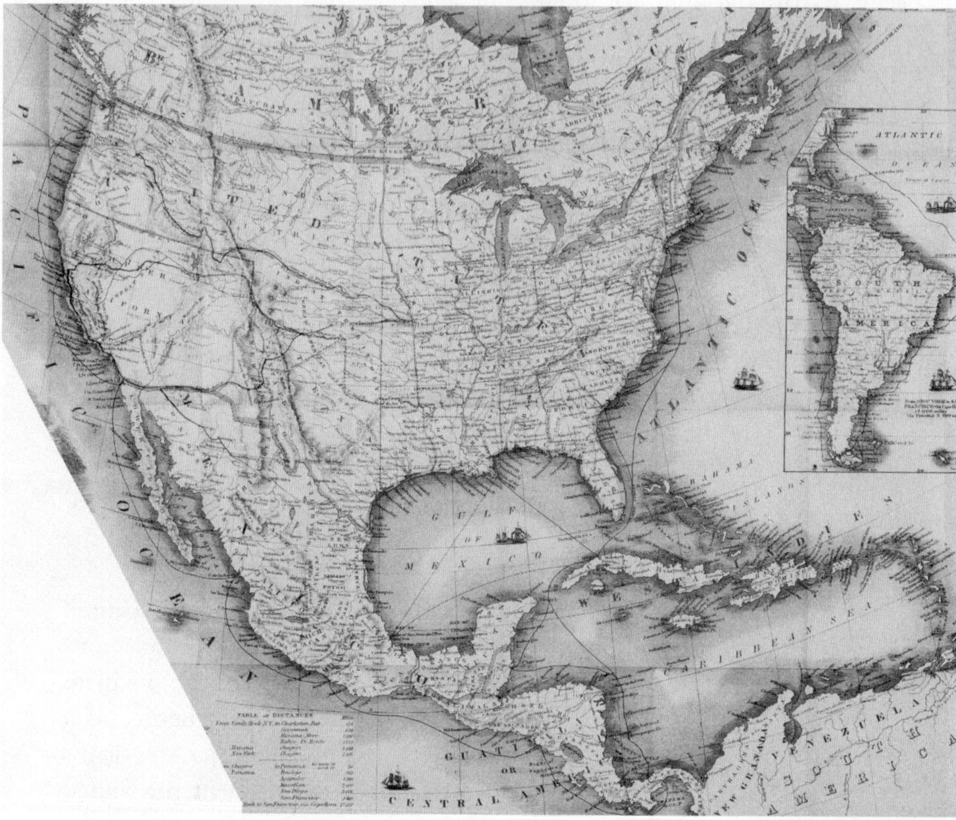

United States Hemispheric Ambitions. From Joseph Hutchins Colton, *Colton's map of the United States, Mexico, &c.* (New York: J. H. Colton, 1849). Courtesy, American Antiquarian Society

[6]

Industrial Manifest Destiny

Eighty years after independence, Britain remained a problem for the United States—in Mexico, in the Pacific Northwest, in eastern counting houses, and overseas. Americans had acquired more land with the Transcontinental Treaty and had found new outlets for manufactures in South America, but these achievements meant little without military or diplomatic might to back them up. The United States had not conquered its new territory; Britain made competing claims to boundaries, and many Indian groups continued to resist land theft. And while manufacturers were making sales in South America, they still could not outsell British goods.

To make matters worse, it was not just England. As US Minister Edward Everett wrote in 1838, "We've got the threat of a British blockade on one side and Canada, Mexico and the Indians on the other!"[1] During the era of ceaseless westward expansion that became known as "Manifest Destiny," a phrase coined by magazine editor John L. O'Sullivan to advocate the United States' annexation of new territory, the nation engaged in costly and deadly warfare, "peaceful" negotiations for the Oregon Territory, and fierce debates over tariff policies.[2]

"Manifest Destiny" has become shorthand among historians for a whole host of nineteenth-century American themes: progress and civilization, individualism and masculinity, racism and imperialism.[3] Embedded within these aspects of expansion was industrial capitalism.[4] Indeed, the "destiny" of American civilization depended on manufacturers who could produce weapons and clothing to sell, brandish, and wear. Historians have done much to connect eastern markets and western space, but before railroads and industrial agriculture dominated the continent, factory textiles made their way into settler homes, and American guns spread throughout the American interior and across the frontier. The years surrounding the origins of the term "Manifest Destiny" were a transitional period in the history of industrialization, at the crossroads between developments in water and steam power and machine tools, and the rise of electricity and large-scale iron and steel production.[5] These seismic technological shifts were preceded by changes in the relationship between manufacturers and the federal government.

The 1840s marked the state's abandonment of private industry. The Ordnance Department, equipped with increased political power, ceased regularly letting large government contracts to private arms manufacturers in favor of procuring from the federal armories.[6] Meanwhile, new federal legislation ushered in a period of relative free trade that left textile manufacturers to face British imports, unprotected.[7] Political agendas shifted as the link between national security and territorial acquisition solidified in the 1840s; overarching federal aims, however, stayed the same. Security, by whatever means and to whatever ends, was of paramount importance. Manufacturing mattered to the executive branch to the extent that it served these aims. It now mattered less to federal officials whether Americans chose caps made of machine-spun wool or dresses stitched together with calicoes that were made at a domestic manufactory rather than one in Liverpool.

This is why we must step back from interpretations of Manifest Destiny that focus on political parties and regional interests. Many studies of the antebellum era analyze territorial expansion in the context of contestations over slavery and southern power, which has obscured other implications of Manifest Destiny. Despite sectional and partisan debates about the morality and constitutionality of expansion, many Americans supported some aspect of territorial extension. The federal government, regardless of party, used the powers of the state to acquire new territory, strengthening industry in the process. The expansionist policies of President John Tyler's administration illustrate this. Although often portrayed as the archetypal pro-slavery, states' rights politician,

Tyler's nationalist objectives and directives often transcended this characterization.[8] As a young congressman from Virginia, Tyler had identified Great Britain as America's main rival for international commercial dominance; he dedicated his presidency to thwarting British efforts in Asia and on the North American continent.[9]

During the era of Manifest Destiny, the federal government pursued expansion and national security through two channels: diplomatic negotiations and military violence. In general, diplomacy characterized the United States' dealings with Great Britain over border disputes and commercial policies and tended to affect the textile industry.[10] Military violence typified the United States' relations with Native Americans and was the chosen course for acquiring land from Mexico. Military action required firearms, of course, but it also opened up opportunities for the textile industry. In the 1840s, Americans' use of guns and cloth, war and diplomacy, played out in two frontier theaters: Mexico and the Pacific Northwest. In both places, the lines between diplomacy and war, and civilian and military goods, were fuzzy ones.

Oregon, Free Trade, and Textiles

In the early1840s, most exported American cotton manufactures went to South America; a decade later, to China.[11] Latin America had been just the first of new outlets for factory goods. In 1832, President Andrew Jackson, who oversaw the first US military intervention in Asia, boasted that "new markets are opening for our commodities . . . we now enjoy the trade and navigation of the Black Sea, and of all the ports belonging to the Turkish Empire and Asia, on the most perfect equality with all foreign nations."[12] Beginning in the late 1820s, American manufactures exported cotton and woolen fabrics to such distant locales as Cádiz, Trieste, Port au Prince, Callou, and Canton. An American visitor to Turkey in 1831, James Ellsworth De Kay, was surprised and pleased to find that cotton manufactures were selling there, just as they were in South America. Even American carpets, still in their infancy, were starting to replace fancy Turkish rugs in markets in Smyrna.[13] A treaty of commerce and amity between Turkey and the United States gave American merchants the same tariff rates as those of "most-favored nations" Britain, France, and Russia. In general, however, the number of exports remained small, especially in comparison to those of Great Britain.[14]

The notable exception was South America, which Ellsworth used as his point of comparison for American manufactured exports to Turkey, and which absorbed more than three times as many American textiles as any other region.[15]

Manifest Destiny, on the continent and overseas, changed things.[16] Americans started to set their sights on China, by way of Oregon, although neither place was new to the American imagination.[17] The Pacific Northwest had been a hot spot for fisheries and the fur trade since before independence. The famed Lewis and Clark expedition (1804–1806) intended to connect the Atlantic and Pacific Oceans by linking the Columbia River with the Mississippi River system. Explorer Meriwether Lewis recognized that direct access to China was the real prize.[18] The United States sent its first commercial ship to China in 1784, where it traded silver, as well as fur, for Chinese silks, spices, and tea. Trade was difficult at best, though, because under Chinese regulations, all commerce had to go through Canton and faced heavy regulations and tariffs. The only thing of real value that Americans had that Chinese wanted was silver from South American and Mexican mines, and to a lesser extent, North American furs. American merchants actually had better luck in exchanging furs for Chinese goods than did their British counterparts (owing to onerous British trade rules and monopolies), but in general British merchants dominated foreign commerce in Canton. British merchants successfully, and tragically, introduced bulk shipments of opium to Chinese consumers, despite resistance from the Chinese government and from American efforts to compete.[19] This trade led to a war between Britain and China in the late 1830s that ended favorably for the former, and to some extent, for the United States as well. Britain and China signed a treaty that opened up four additional ports, a provision from which Americans would profit. It also granted Britain most-favored-nation status (and the right to fix tariff rates), to the chagrin of American merchants.

The signing of this treaty, on August 29, 1842, came just a week after the ratification of a treaty between Great Britain and the United States, which seemed to resolve a whole host of grievances that had plagued the two nations for decades. US Secretary of State Daniel Webster and British minister Alexander Baring, 1st Baron Ashburton, worked out a deal that settled border disputes in Maine and the Great Lakes region, and committed to a final suppression of the coastal slave trade. The treaty marked an achievement in US relations with Britain not only because it solved a number of long-standing problems, but also because Webster and Ashburton negotiated on fairly equal and amicable terms.[20] President John Tyler said that "the treaty lately concluded with Great Britain has tended greatly to increase the good understanding which a reciprocity of interest is calculated to encourage, and it is most ardently to be hoped that nothing may transpire to interrupt the relations of amity which it is so obviously the policy of both nations to cultivate."[21]

The treaty, however, did not resolve the problems resulting from an uneasy joint occupation of the Pacific Northwest. Since 1818 Americans and British agreed to coexist in Oregon, which worked well enough for the ensuing two decades, but not without occasional problems. Britain controlled the fur trade, and its Hudson's Bay Company made things particularly difficult for American settlers by charging exorbitant amounts for land and interfering with business operations.[22] The two diplomats' decision to table the Oregon issue came at a transitional moment in US interest in the Pacific Northwest. After the colonization enthusiasm of the 1820s died down, settler and missionary activity picked up again in the 1830s and many settlers sought US governance.[23] American citizens from all over the country sent petitions to Washington for financial and military aid to settle Oregon, making appeals to agricultural opportunity, poor treatment at the hands of the British, and the need for a bulwark against foreign interference.[24] Additionally, Pacific whaling was on the rise, and the region, after a recent publicly funded exploratory expedition, was starting to gain notice for mineral resources in the area.[25] Despite Oregon's growing presence in national discussions, the Webster-Ashburton Treaty, which was silent on the issue, met little resistance and was signed and ratified in less than two weeks.

The achievements of the treaty, and Webster's and Ashburton's decision to postpone the Oregon issue for the sake of consensus on other issues, did little to abate competition between Britain and the United States. When China and Britain signed the Treaty of Nanking at the end of that month, the opportunities and challenges were not lost on Americans, especially in regard to Oregon. A House committee on the practicality of settling Oregon reported on "the prospect of the opening of the ports of China to the commerce of the world, as they now are to Great Britain" and predicted that "our enterprising republican people in Oregon could outstrip all other competitors in availing themselves of those interests."[26] The commercial prospects Oregon offered mattered, because in Canton US merchants complained about disrespect and claimed that fluctuating tariff rates hampered their ability to compete with British merchandise, which increasingly included textile imports.

Up through the 1810s American merchants shipped furs, ginseng, and sandalwood, in addition to silver, to China, a commerce that yielded a potential 200 to 500 percent return on a two-year voyage. They also profited from their status as neutrals in the European re-export trade. The return of French and Dutch merchants to China after the Napoleonic Wars, however, cut into Americans' European re-export trade. Additionally, fur, which was being depleted in North America, could be difficult to profit from, especially because US demand for tea so far exceeded Chinese demand for furs.[27]

Enter textiles. China had a long history of textile production and its consumption of textiles was similar to England's (about eight pounds per person).[28] Increasingly, however, Chinese consumers were warming up to foreign cotton manufactures. As British and American merchants imported cloth woven on power looms, wealthy urban residents looked to fine lightweight cottons as substitutes for silk fabrics, while poor laborers purchased coarse fabrics that were more durable for work.[29] During the 1820s and 1830s, when American textile factories were finally starting to become profitable, the United States exported up to $500,000 worth of domestic white cottons and $65,000 worth of printed and colored cottons—enough to make Britain nervous, even though Britain's trade still exceeded that of the United States. In 1830, for example, Britain sold about $143,000 worth of its cotton manufactures in China, while that same year the United States' domestic cotton exports to China only equaled about $56,000 (this was a low year for both nations).[30] Many Americans believed this disparity in trade resulted from unfair arrangements between the British and Chinese, especially after the Treaty of Nanking. Just as in South America, Americans thought that with a "fair chance" their goods could outperform British ones because they were of higher quality.[31] One company that was particularly successful with exports and had gained national recognition was the Hamilton Manufacturing Company in Lowell, Massachusetts. Its sales showed what a wide geographic range was possible for American textile exports: in the late 1820s, the company shipped to Cincinnati, New Orleans, Smyrna, Buenos Aires, and Malta.[32] Soon after, they also shipped to Canton.[33] Still, increased British power in China threatened to cut into American profits. To compete effectively in China, textile manufacturers needed diplomacy, just as they did in South America.

The notion of competition is somewhat misleading. Many textile manufacturers had conflicting economic and moral interests, which made it difficult to identify a clear competitor or competitive strategy. Take George Howe, for example, who had decades of shipping experience in the Pacific Northwest and China.[34] He did business with London merchants to whom he was both a debtor and a creditor.[35] As agent, treasurer, and director of various New England textile factories, he relied on southern plantation owners—who generally opposed protective tariffs on manufactures—for cotton.[36] Many of the firms Howe worked for and with belonged to some of the same men who had benefited from the claims payouts of the Transcontinental Treaty and were now owners of the most successful northern factories. These elite textile manufacturers derived their wealth from varied and sometimes illicit international sources; they had jettisoned moral antislavery in the service of political and economic gain. Historians of the nineteenth century United States have rightfully paid much attention to this reality.[37]

There was something bigger happening, though, beyond complicated and shifting alliances and concerns. President Tyler decided that American diplomats should negotiate for trade terms that would allow the United States to regulate international tariffs, even though only a small contingency wanted an envoy sent to China. Some Americans saw the China trade as a bastion of privileged elites and moral depravity, especially because a portion of that trade involved opium.[38] Others simply reasoned that the trade had been going on successfully enough for decades and there was no sense wasting national resources on a diplomatic mission.[39] Nevertheless, John Quincy Adams, now a representative from Massachusetts, pushed through a $40,000 appropriation in the House for a diplomatic mission to China in 1843. Caleb Cushing, a congressman from Massachusetts whose political constituents, family, and friends were experienced China traders and textile manufacturers, was chosen to negotiate a treaty with the Chinese emperor.[40] The circumstances surrounding the mission looked fishy to critics. Just before he resigned as secretary of state, Daniel Webster had appointed Cushing to the post of negotiator. Webster owed Cushing (along with other influential Bostonians) money, and Cushing himself was related to one of the wealthiest

China traders, John Perkins Cushing. Not surprisingly, Webster ignored the wishes of the mission's opponents, who believed that any negotiations should involve a guarantee of US naval force to suppress the coastal opium trade. Instead, he declined to give Cushing any instructions at all related to that issue; and so, the opium trade could carry on, free from regulations.[41]

This squabbling was just part of the story, and ultimately a minor part. A major objective of the treaty was to secure most-favored-nation status, just as Britain had done. After Cushing arrived in China, however, he learned from the US minister in London that this status already had been achieved by a supplement to the Treaty of Nanking a year after its signing, which put the United States and several European nations on equal footing. The appearance of US diplomatic strength, however, was important to Cushing and the Department of State; it was decided that he stay and negotiate anyway. The Treaty of Wanghia passed the Senate by a unanimous vote on January 16, 1845, overcoming prior opposition to the mission.[42]

One "achievement" of the treaty was the ability to influence tariff rates on American goods, something that mattered a great deal for textiles imported into China. Skeptics of the treaty were right to be critical of the opium connection, but many China trade merchants had recently started investing in textile factories. This included John Perkins Cushing, who invested almost one-third of his commercial holdings into textile manufacturing in the 1830s; it also included his uncle, Thomas Handasyd Perkins, who traded furs and opium throughout the 1820s and 1830s, but poured increasing amounts of his wealth into some of the Boston Associates' various undertakings.[43] The treaty with China mattered for smaller factory owners, too, who were starting to list exports on census questionnaires that asked about where their goods sold. Many manufacturers were not sure exactly how many of their goods were going where, but after the treaty took effect, the value of cotton manufactures that the United States exported to China doubled from $813,606 in 1846 to $1,691,959 in 1848.[44]

Two other things were happening at the same time that influenced these numbers. One was that the fervor for consolidating US control over Oregon reached its zenith shortly after the signing of the Treaty of Wanghia. A Senate committee on postal roads exemplified the heightened anxiety over the Pacific Northwest; in its 1846 report on the expediency of establishing different postal routes, the committee veered off topic from postal communication to argue that not taking enough land in the Northwest was tantamount to "infanticide." It warned of Britain's attempts to "crush" American commerce. Oregon, according to the committee's logic, was important not just for furs or fish, but for squashing imperial rivalries and monopolizing the trade with Asia.[45] The views expressed in this report were a more aggressive iteration of previous political inclinations that connected the conquest of Oregon with improved Asian commerce. Several years earlier, for example, a House committee on the expediency of settling Oregon included in its report the observation that "opening the ports of the Celestial Empire to the commerce of the world will tend greatly to enhance the value of our interests in the pacific."[46] Oregon's value was thus inseparable from its potential for facilitating Asian trade. Another committee advocated for military posts in and on the route to Oregon to institute a kind of standing police; otherwise, US claims to the territory would remain "barren or untenable."[47]

James K. Polk won the 1844 presidential campaign on a platform that included claiming all of Oregon for the United States, and by 1846 the jingoist phrase "Fifty-four Forty or Fight" had become a rallying cry for expansionists.[48] In the spring of 1846, Congress passed a joint resolution to end shared occupation of Oregon and to negotiate peacefully with Britain.[49] British diplomat Richard Pakenham and US Secretary of State James Buchanan signed an agreement that gave the United States all of the Pacific Northwest south of the forty-ninth parallel. This was less than what fervent expansionists wanted, but it was more than enough to resolve territorial disputes and to provide the United States adequate maritime access and customs revenue. It also pacified relations with Britain so that the United States could focus on war with Mexico, which Congress declared on May 13, 1846.

While this version of Manifest Destiny is well known, jingoist entitlement to land on the Pacific coast is not usually associated with eastern textile mills. But Oregon was connected to the China trade, which increasingly included items like cotton drillings and jeans. Any improvement in access to Asian markets was useful for manufacturing—indeed the State Department singled out merchants from New York and Massachusetts as the primary beneficiaries of improved communication between the Pacific coast and China.[50] Once the Oregon Treaty was ratified, railroad boosters sought federal support to connect eastern markets with Oregon. Their argument centered on the fact that improved transportation would enable the United States to compete with foreign merchants because it would lower the price of goods and thus increase Pacific demand for woolen and cotton manufactures, which already "formed a large item in that trade."[51] All of these arguments recognized that American manufacturing depended on markets, transportation, and new land.

The improvement of access to markets was especially important because the short-lived era of high protectionism was coming to an end. At the same time American expansionists clamored for Oregon, lawmakers debated major changes to the nation's trade policies. From 1842 to 1846, American manufacturers enjoyed a tariff schedule that taxed textile imports at 70 percent. Most manufacturers prospered under the protective tariff. When the Treasury Department solicited responses to questions about local economic conditions, the Oneida Manufacturing Society answered that area mills had seen profits of over 12 percent under the 1842 tariff (more successful firms like those of the Boston Associates were higher, up to 19 percent), in contrast to the dismal yields before the tariff was enacted.[52] Legislatures from states with a lot of manufacturing sent petitions to Congress, asking federal lawmakers to maintain the tariff levels of 1842.[53] They did not listen. Secretary of Treasury Robert Walker proposed a plan in 1845 to dismantle the protective system and reduce the tariff on cotton and woolen manufactures to 25 percent.[54] Part of the rationale behind tariff reductions was congeniality with Britain, because Parliament concurrently was debating changes in commercial policy that would benefit American grain exporters. Led by Prime Minister Robert Peel, Britain repealed its prohibitive tariffs on grain imports in May. Two months later Congress reciprocated by passing into law a tariff bill based on Walker's suggestions. American proponents of tariff reductions celebrated Peel's efforts and acknowledged that the US tariff system was founded on Britain's and should be changed accordingly, especially if Americans hoped to take advantage of more open British markets. There was significant debate over what to do about tariff rates, but war with Mexico broke out in April 1846, and in a Democrat-controlled Congress that largely favored territorial expansion, revenue for war took precedence over all else.[55] Representative Seaborn Jones from Georgia argued that protective tariffs were prohibitive and deprived the nation of war-making revenue. Besides, he reasoned, "eastern manufactures are carried to such an extent now—to such skill and perfection, both in men and machinery—that they can compete with the manufactures of Europe."[56]

Jones was actually right. For the rest of the decade, textile manufacturers complained about the 1846 tariff, but they did not actually curtail production.[57] New markets in China and better access to the Pacific helped temper the "free-trade" policies of the mid-1840s. From 1840 to 1850, the quantity of cotton manufactures almost tripled.[58] Then, even as some textile manufacturers bemoaned the expansion of slavery, they got an added boost when the United States went to war with Mexico. The sum total of competing interests was an expansion of territory and industry.

The Mexican Frontier, War, and Guns

At first, Joel Roberts Poinsett did not think war against Mexico was a good idea. A potential defeat against a "weaker nation" bothered the former secretary of war, especially since the United States had only just recently stalemated in a humiliating war against the Seminole Indians in Florida.[59] Throughout the spring of 1846, the thought kept him up at night.[60] He was not, at least, concerned about the nation's firearms. Indeed, a Mexican officer visited the United States in the early 1840s to observe its first-rate artillery, even as Mexico was able to import plenty of British weapons.[61]

Arms production had improved since the 1830s, when insecurities about how US military production stacked up against Europe plagued federal officials. The US government continued to import some of the guns used as gifts for Indians, and those it contracted for domestically had to be modeled after the British "northwest" rifle—a light firearm used in the fur trade.[62] In the 1830s, diplomats and army officers still looked anxiously overseas. US minister Richard Rush cautioned the War Department that "we live in an age when the world is moving forward . . . the French have made improvements in guns."[63] Similarly, when Poinsett served as secretary of war, he received a letter warning him that "we do not value mechanical and manufacturing industry enough."[64] In an evaluation of a new breechloader's susceptibility to explosion, the Ordnance Board reported that "these objections may be overcome by those in Europe who are devoting great attention and consideration to this [style of gun]; if so, we should place ourselves on a footing with those nations who may adopt it, and to whom hereafter we may be opposed."[65] Several years later, the federal government subsidized a trip to cannon foundries, small arms manufactories, and arsenals in Europe. As ordnance officials toured England, Scotland, Sweden, Russia, Prussia, Belgium, and France, their anxieties faded. They returned with the satisfaction that once they replaced all flintlocks with percussion locks, their arms would fire reliably in any weather—and would be superior to arms manufactured in Europe.[66]

It was increasingly becoming the case that the United States, not Europe, was the hub of arms making. The exceptions were specialized hunting guns, which even some military officers purchased from England, despite owning American-made pistols, rifles, and muskets.[67] This reflected the extent to which American arms making was devoted to military, not civilian, production. Europeans had begun taking notice of American military firearms and sending their own officials to visit US armories. Most improvements in gun production occurred at one of the national armories or at the factory of government contractors, who remained based in Middletown and New Haven, Connecticut, and in Pittsfield and Millbury, Massachusetts. Asa Waters of Millbury, Lemuel Pomeroy of Pittsfield, and Eli Whitney of New Haven produced muskets on contract, and Nathan Starr, Robert and John D. Johnson, and Simeon North, of Middletown, manufactured swords and rifles, rifles, and pistols, respectively. From the 1810s onward, manufacturers like Nathan Starr acknowledged that the operation of a "large and expensive factory" depended on "steady encouragement from the government."[68]

Although the federal government cultivated its small cohort of private arms contractors with cash advances, it had begun experimenting with other types of relationships in the late 1810s. Thomas Blanchard was a member of the Connecticut Valley network of engineers and manufacturers whose skills the Ordnance Department sought, but who did not have his own gun factory in which the government could invest. Blanchard, of Sutton, Massachusetts, was an inventor, whose upbringing was in many ways emblematic of the idealized American success story. Born to a poor family and beleaguered by a stutter and minimal schooling, he eventually became, according to a popular 1863 cartoon, a "man of progress."[69] Over the course of his life, he patented over fifty inventions, the most important of which was a lathe for turning gun barrels. The New England legend that "the idea for this machine came to [Blanchard] it is said by watching the labors of a wood-boring insect" may or may not have been true. Regardless, the lathe's implementation in interchangeable production required federal support.[70]

In exchange for the use of Blanchard's inventions, the Ordnance Department provided Blanchard tools and machinery and paid for his work at the federal armories. The relationship began in 1818, when Asa Waters invited Blanchard, who had a reputation as an innovative mechanic, to his factory in Millbury, Massachusetts. Waters had invented a lathe to turn gun barrels, but had not figured out how to eliminate the need for hand filing, so he called on Blanchard for help.[71] Blanchard integrated filing capability into Waters' barrel-turning process by adding a movement of the cam, an accessory that adjusts the path of the material to create a non-circular curve.[72] News spread quickly among the ordnance network and Roswell Lee invited Blanchard to the Springfield Armory to demonstrate.[73] After tweaking the armory's lathes to include his cam motion, Blanchard went home to attempt a machine for making gunstocks, which were extraordinarily labor-intensive. Even a skilled group of workers could only turn out eight or ten in a week.[74] When Blanchard successfully completed a miniature version, Lee proposed to the Ordnance Department that it hire Blanchard, so that Blanchard could perfect his stocking machines using the armory's resources. In 1822, Lee and Blanchard agreed that Blanchard would receive thirty-seven cents per musket and free use of armory tools, and in exchange, Blanchard would hire his own workers and allow the armory free use of his patented lathe.[75] By the time Blanchard left public employment in 1827, he had received $18,500 in patent fees.[76] On the government side, historian Carolyn Cooper estimates that Blanchard's half-stocking machines delivered a net savings of $47,340 at Springfield Armory and a total of $89,239 at both federal armories between 1820 and 1839.[77] The relationship, however, did not last as long as did those with regular contractors: when Lee decided that the armory had obtained all the benefits it could from Blanchard's work, Ordnance terminated his employment.

In the 1830s, it became more likely for innovations like Blanchard's to occur outside of public sponsorship. Federal investment had nurtured the industry so that others could build off decades of developments without taking on the burden of research costs. The legendary Samuel Colt, for example, had his start in the 1830s when he applied US and British revolving techniques to rifles and pistols. Colt's father, a textile manufacturer from Ware, Massachusetts, funded his early arms making before Colt traveled to England and Scotland in 1835 to secure patent rights. The early US patent system was notoriously problematic for making a profit, which was partly why Blanchard shared his patent rights with the US government. In 1836, however, the system was reformed, making patents harder to acquire and thus more valuable. These reforms reflected a shift away from the encouragement of invention for the sake of invention, and toward the protection of industrial capital, which made patents more appealing for non-government investors.[78] One of Francis Cabot Lowell Jr.'s business associates, for example, specifically sought out John W. Cochran, an inventor who manufactured rifles at a private factory in Springfield, for the opportunity to invest in the manufacture of his "celebrated rifle."[79] Cochran and Colt both received patents under the new system; both embodied the new class of private firearms manufacturers.

In 1837, the Senate required the War Department to conduct an examination of the improvements in firearms made by non-contractors.[80] Colt's and Cochran's firearms, along with those of John H. Hall, Daniel Leavitt, and Baron Hackett, were tested in the examination, which occurred amidst a general climate of government reform and cost effectiveness in the 1830s.[81] "Cost effective" did not necessarily mean "private sector." In fact, the arms made by John Hall, the only manufacturer who had worked for the government, fared best in the examination.[82] In general, the War Department was wary of inventors like Colt, who were motivated by profit rather than battlefield realities. It was not that War Department officials did not value innovation. When Joel Poinsett served as secretary of war, for example, he invited policy makers and inventors to "come and talk about guns" in order to learn about new inventions.[83] The War Department, however, valued military applicability over novelty, and tended to err on the side of safety and reliability. This was why the ordnance examiners ruled against private arms makers like Colt, justifying their evaluation by saying that "in the desire to reach perfection in the construction of these arms, the convenience, physical power, and safety of those who are to use them have been wholly overlooked."[84]

The examiners further determined that "the arms now in use in service of the United States . . . combine all the requisites of convenience, durability, simplicity, and efficiency."[85] Although the evaluation did not result in a definitive conclusion about whether privately produced guns were better than public ones, it was a sign of the changing relationship between private and public production. By the 1830s, the federal armories produced about 80 percent of the nation's serviceable arms, which meant that the government no longer needed to sustain long-term relationships with manufacturers. It issued the last round of advance-sum contracts in 1834 and began altering contract terms to reap maximum military benefit at minimum public cost.[86] The Ordnance Department now reserved the right to nullify the entire contract if more than three-quarters of it were not filled, or if, in some cases, like Lemuel Pomeroy's 1840 contract for six thousand muskets, 100 percent of the annual amount was not met.[87]

This caused anxiety among regular contractors, who, unlike Colt, had dedicated their entire careers to government manufacture. Asa Waters, who had spent over a quarter-century making muskets and pistols for the government under contract in Millbury, Massachusetts, begged ordnance chief George Bomford for additional work in 1841, promising to produce pistols 10 percent more cheaply than could be done at the national armories. He even offered to forgo payment for over a year if that would be more amenable to the ordnance budget.[88] Waters spent the next few years looking elsewhere for business, but he never stopped applying for government work, even though, as Waters said to another contractor, "they keep applying the screws closer and closer to grinding harder upon the contractors."[89]

The federal government was able to be more exacting with its regular contractors partly because of competition from private arms makers.[90] One year after the 1837 Ordnance exam, the Committee on Military Affairs recommended that the government purchase Hall's patent rights.[91] That same year, Colt got his first government business when he sold five hundred rifles to Quartermaster General Thomas Jesup in Florida.[92] Although Colt's first revolvers, which he had patented in 1836, were impractical for field use, his improved versions had the potential to permit US troops to fend off a Seminole offensive. They fired more than ten rounds in a minute, and their ramrods, which many men dropped in the loading process, were attached to the body of the weapon. Additionally, Colt implemented a loading lever so that the hammer rested on a safety pin situated between two caps, rather than on the cap itself, to prevent the weapon from firing unintentionally.[93]

Again, however, officials wanted to test them.[94] On November 18, 1840, a board of officers of the first dragoons met in Pennsylvania to compare Colt's repeating firearms to Hall's carbines, for which he had secured a patent several years earlier. The board conducted ten experiments, an example of which involved the carbines "slung to a man mounted, who galloped rapidly for a mile, the piece swinging against the side of the horse." Colt's carbines held up well to rough use by the experimenters and were faster than Hall's—firing eighteen rounds in two minutes, forty-five seconds, compared to Hall's eight minutes. They were less accurate, however, because although smoothbore barrels increased speed, their loads did not fit as tightly, which could decrease accuracy beyond short distances. Hall's carbines hit the target eighty-seven times, to Colt's sixty-nine. Ultimately, the board reported that "foregoing experiments were very successfully made, and have impressed us with the belief of the utility of these repeating fire-arms for military purposes." They tempered their optimism with caution. Because of "the deafening sharpness of the report, which must injure the hearing of those who use them," the board recommended a six-month trial in the field and garrison.[95] Colt received a contract for two hundred carbines the following year.[96] As the government started to do business with new private arms makers, it could be selective about the terms because it was still the biggest consumer.[97] Although Samuel Colt would eventually become one of the nation's most successful arms manufacturers, he claimed that prior to the Mexican-American War there was not a civilian market for revolvers. Colt's attorney attributed his firm's 1842 closure to the fact that "the moment the Indians [in Florida] were extirpated, there was no market for their guns."[98]

The government also shored up its control over the public sector by replacing the civilian superintendents of the federal armories with ordnance officers. West Point–trained managers used new accounting methods to make more efficient internal production processes.[99] New Englanders, far removed from the frontier and military practicalities, resented this change. Springfield employees and town residents, for example, petitioned Congress to avoid changes in the law. Changes, they argued, "may be proper in the organization of the army and navy, but are degrading, oppressive, and tyrannical when applied to intelligent and high-minded citizen mechanics."[100] Civilian arguments, however, were no match for War Department goals to appoint men with military experience to oversee weapon production.[101] Civilian James Robb was replaced by James Ripley, a major of ordnance who had fought in the Seminole Wars.[102] Some historians have described Ripley as averse to innovation, based on his suspicion of weapons makers like Colt, and his slowness to adopt new technologies, but earlier in his career, Ripley was responsible for improvements to artillery. Also, innovation for innovation's sake was not always best for security; Ripley's caution likely reflected his experience with military operations.[103] Either way, his appointment so infuriated workers in Springfield that they brought a lawsuit against him, accusing him of unfairly laying people off, wasting public resources, and causing a deterioration in the quality of arms.[104] Just as their petitions against changes in superintendence failed, so too did this lawsuit. The government kept Ripley on, and despite his management style, the Ordnance Department was able to meet requests for additional supplies and arm their troops with mostly new guns. During Ripley's tenure, the Model 1816 (flintlock firearm, infantry musket) was replaced with the first conventional musket of interchangeable parts: the Springfield Model 1842.[105]

This development was well timed for the Mexican War. The Mexican frontier had long plagued American policy makers and once Texas became an independent republic in 1836, Americans debated admitting it to the union, along with other Mexican territory to which they had dubious claims.[106] Insecurities about Britain, which urged Texas to maintain its sovereignty rather than join the United States, drove much of the discussion. Many Americans argued that if the United States did not annex Texas, Britain would foment Indian and slave rebellions. President Tyler had long identified Great Britain as America's chief rival in its quest for global commercial supremacy.[107] He and pro-slavery expansionists argued that England would prohibit the spread of slavery, but even northern opponents of Tyler's administration argued for annexation because they feared that British presence would stymie trade.[108] France was also a problem. Although the United States had supported France in a minor war against the British-backed Mexican government in the 1830s, France, like Britain, opposed annexation. The French government wanted unfettered access to Texan markets and feared the geopolitical consequences of an expansive United States.[109]

Secretary of State John C. Calhoun completed an annexation treaty with Texas on April 12, 1844, and Mexico immediately severed diplomatic ties with the United States. The US Senate defeated ratification, but annexation remained a major issue in the 1844 election. James K. Polk won the presidency on an expansionist platform, and the US Congress made Texas a state in December 1845. It declared war on Mexico less than six months later. These decisions marked a total abandonment of "security" policies in favor of martial belligerence. This was a war for territory, and in order to wage it, the federal government gave up protective tariffs and negotiations with Britain for more of the Oregon Territory. Historians ascribe varying degrees of bellicosity to Polk, whose presidency was dominated by conflict with Mexico. Some say he singlehandedly orchestrated the war, while others claim that a variety of policy mistakes beyond his control led to hostilities.[110] Regardless of his individual culpability, the war revealed an executive commitment to martial aggression.

When the United States declared war on May 13, 1846, it was the first time it had done so since the War of 1812. Like that earlier war, there was a strong anti-war contingent; unlike it, Congress voted decisively in favor of war. In Polk's request to declare war on Mexico, he argued that US troops needed reinforcement after Mexican troops unjustly attacked them on American soil; in reality, General Zachary Taylor helped instigate the conflict by invading Mexican territory. Democratic war hawks attached the declaration of war to a bill that authorized funds for the troops, making a vote against the bill a vote against the troops. It passed the Senate forty-two to two, and only fourteen members of the House of Representatives voted against it.[111]

This was perhaps the first time the United States was prepared for war. Gone were the insecurities over weapon supply. As one Philadelphia-area newspaper noted, the government had plenty of "muskets ready for shipment at a moment's notice. There need not, therefore, be much fear of a scarcity of guns."[112] Two months after fighting commenced, the War Department reported that the number of arms produced at Springfield greatly exceeded that of the previous year.[113] By June of the following year, the United States had over $8.4 million worth of small arms in its inventory.[114] Many of these guns represented the latest in firearm technology, including the first conventional musket made entirely of interchangeable parts.[115]

This newfound weapon security was the result of financial priorities. During discussions over wartime funding in June 1846, ordnance expenses for the first month of combat were expected to total $603,000. An amount of $248,000 was predicted for the subsequent fiscal year (July 1, 1846, to June 30, 1847).[116] Secretary of Treasury Robert J. Walker reported that the aggregate War and Navy Department expenditures for the fiscal years ending on June 30, 1846, and June 30, 1847, would exceed by $24 million the amount projected the previous December. Walker argued, though, that if the proposed tariff bill—which virtually abolished protectionism—passed, it would increase trade and produce a net revenue of $3 million.[117] This sort of argument marked the executive's shift from industrial security to aggressive militarism, a sacrificing of small producers for war. The eradication of protectionist policies was a slam to textile manufacturers, most of whom believed they depended on tariffs for their profits, in contrast to arms contractors, who knew that direct federal investment, rather than tariff policies, dictated their success. In the words of arms contractor Asa Waters, "If the patronage of the government is not continued, our factories will be worth but little."[118]

Walker's privileging of military funding over industrial security had actually already begun permeating the arms industry as well. Instead of building up the factories of private arms makers, the government began altering contract terms to reap maximum military benefit at minimum public cost. In its 1839 contract with Simeon North for ten thousand carbines, for example, the Ordnance Department expected "perfect uniformity," and to be able to "exchange parts without impairing efficiency," but reserved "the right to annul any part of the contract."[119] This was a stark change from the 1810s and 1820s, when ordnance officials had negotiated extra funding for North's manufacturing pursuits.

The distancing between contractors and federal officials only increased during the war. The Springfield Model 1842, for example, which represented the labors of both federal armory employees and individual contractors, became almost solely the product of the national armories. Simeon North and John Hall had been largely responsible for introducing the percussion lock system to US carbines and rifles and for developing milling machines that made possible the manufacture of interchangeable parts; by 1846, however, only a handful of this model were produced outside of the federal armories. And so, even with the almost $1 million it received during the war, Ordnance became less generous with contractors and instead spent the money to improve infrastructure at the federal armories and arsenals and to update its machine-tool inventory, which included machines for introducing percussion cap technology to flintlock muskets and rifles.[120]

In many ways Ordnance was capitalizing on decades of direct investment in private armories; it now used federal resources to consolidate control over production. Since the 1820s, several contractors owned more sophisticated milling machines and employed more skilled artificers than the federal armories.[121] North and nearby manufacturers Robert Johnson and Nathan Starr employed some of the best machinists in the country, whose wages were paid with government capital. Springfield superintendents Roswell Lee and John Robb dispatched armory employees to these contractors' factories to observe mechanical processes.[122] They also began enticing these workers to Springfield. After contractor Lemuel Pomeroy paid for Scottish engineer Henry Burden to introduce new waterwheel technology at his factory in Pittsfield, Massachusetts, Lee brought Burden to work at Springfield.[123] The Springfield Armory began to absorb much of the mechanical talent in the region, and by the 1840s, the quality of its employees was unparalleled.[124]

As the federal armories appropriated the labor and mechanical improvements that had been made over the previous decades, the Ordnance Department turned to shorter-term agreements—solicited through advertisements—even as it recognized that the "system of advertising would result in the final ruin" of some of its regular contractors.[125] This represented a sort of convergence in the government's approach to the arms and textile industries. Just as the federal executive had always kept its distance from textile producers, it started to behave the same way with non-armory production of weapons. During the Mexican-American War, the Ordnance Department had little need to enter into new contracts; for the most part it finished collecting on preexisting contracts. It did, however, purchase two thousand pistols from Samuel Colt during the war, an arrangement that exemplified the benefits of its new approach to the private sector. Unlike earlier contractors, Colt, not the federal government, had to bear the expense of producing arms for the US Army. Although War Department officials were originally skeptical of the new revolving technology used in Colt's patented firearms, the Texas Rangers used Colt's revolvers successfully against Mexican troops in the early 1840s, and Captain Samuel Walker heartily endorsed them to the Ordnance Department. When the department issued an order for one thousand pistols early in the war, Colt's first firearms factory had been closed for three years, and he had to subcontract the work to Eli Whitney Jr. Whitney claimed he was unfairly compensated—and Colt barely broke even. The second order cost Colt $30,000 for tools and machinery, but after he received $28,000 for the one thousand pistols, Colt was $2,000 in debt.[126] From the government's standpoint, this arrangement worked particularly well. Colt bore most manufacturing costs, while US troops had success with his revolvers, which received glowing reports from the battlefield. D. E. Twiggs, Seminole War veteran and commander under both General Zachary Taylor and General Winfield Scott, endorsed them to Congress, which made the Ordnance Department's decision making look good.[127]

In general, the War Department's modified approach to supplying the army with small arms was a successful one. The biggest problem facing Americans in Mexico was manpower, as Poinsett had warned. At the start of the war, Mexico had a standing army of about thirty-six thousand, compared to just over seven thousand men in the US armed forces, fewer than before the War of 1812.[128] American officers repeatedly wrote about the superiority of the enemy's numbers. They also wrote about the efficacy of American small arms. The Battle of Buena Vista in February 1847 was a US victory that has largely been attributed to the superiority of American artillery over Mexican troop numbers. That winter, however, it was not just recently developed mountain howitzers (artillery pieces that fired shells over high trajectories) that enabled American soldiers to kill more Mexican soldiers than the number of Americans killed by Mexicans; it was also rifles, pistols, and muskets. Colonel Humphrey Marshall of the Kentucky cavalry reported that his regiment of four hundred, "armed with rifles, or with carbine, pistol and sabre," was successful against almost fifteen hundred men. Another commander noted that "notwithstanding the great superiority of their numbers, [our] riflemen kept up a deliberate and well-directed fire upon them," and General Zachary Taylor boasted that Americans "maintained their ground handsomely against a greatly superior force, holding themselves under cover and using their weapons with deadly effect."[129]

However unpopular the war was, for many civilians on the home front there was a perverse comfort in the nation's ability to supply troops readily with guns and ammunition. One newspaper reported several days after the declaration of war that "we learn that over 2,000 muskets and over 700 kegs of ball and buck shot cartridges . . . are destined for the Rio Grande. Our troops will no doubt 'sprinkle' the latter liberally upon the Mexicans."[130] The cartridges were the same kind that had been used to kill Seminoles in Florida.

1848

Negotiators from the United States and Mexico signed a treaty of peace on February 2, 1848. After capturing the capital, the United States government had to decide whether to push for all, or just some, of Mexico. No matter what Polk or aggressive expansionists wanted, American troops just wanted to go home. John C. Calhoun, who had become an ardent states' rights proponent since his days strengthening the War Department and federal power, opposed absorbing Mexico on the grounds that the United States was "the government of the white man." (He was pleased with how the military had performed, however).[131] Instead of annexing Mexico, the United States agreed to the Rio Grande as the boundary between the two nations.[132] Mexico forfeited large portions of the Pacific coast south of Oregon and a sizable chunk of the territory west of Texas.[133]

The United States acquired California just in time to benefit from the "discovery" of gold by a sawmill operator in the Sierra Nevada. But while gold was the glitzy prize, Americans also got superior harbor access to the Pacific, which would bolster its trade with Asia. California was an excellent complement to Oregon, which became a US territory in August of that year.[134] Two years after the United States secured its control of disputed land in the Pacific Northwest, American petitioners requested official US governance in Oregon, once again linking military security and commerce. They sought federal resources to make the Columbia River more traversable for whalers and merchants and argued that the foodstuffs produced in the region would feed land troops and the Pacific navy. Once adequately governed, petitioners said, the territory would provide unparalleled access to Asian markets and security against enemies who attempted to enter the United States from the Pacific.[135]

Expansionists were right to connect economic and military benefits. All the violence and diplomacy of the previous decades was a boon for industrial growth. Even as Asa Waters bemoaned the decrease in government contracts for arms makers, he commented on the unprecedented growth of cotton manufacturing after visiting the Boott Mills in Lowell: "It has seemed almost impossible for the increase of supply to keep pace with the increase of demand. Especially since the opening of the China market for Americans cotton, so large proportion of those manufactures are exported."[136] By the early 1850s, American tonnage to China far surpassed that of all other nations except Britain. After Oregon became a territory, the export of jeans from the United States to Canton more than doubled, from fourteen thousand pairs in 1848 to thirty-two thousand in 1850 and forty thousand in 1851.[137] And once the United States acquired California, its trade with Chile, for example, surpassed England's. American-made cotton textiles were a particular point of pride in the US State Department's commercial report.[138] The domestic market continued to absorb about 90 percent of US manufactures, but new outlets helped manufacturers withstand changes in commercial policy, such as the removal of protectionist tariffs in 1846.[139] Ironically, in between the start of war with Mexico and the passage of the Walker tariff, one newspaper writer commented that "the government now sees the advantage of a protective tariff that enables them to procure at low rates and at home the necessaries of life for the soldiers."[140] Protectionism died, but the writer was not wrong. By the 1840s, the army was supplied regularly with clothing of domestic manufacture.[141] Cotton manufactures exported to Mexico increased over fifteen times throughout the war and in the years after remained higher than they had been before the war.[142] Textile production had caught up with gun production in terms of quantity and quality. It, too, could now survive both war and peace.

Neither industry, however, ever really had to withstand peace, because warfare was endemic to Manifest Destiny. Troops were now "almost constantly in the field" in the new territories. Quartermaster General Thomas Sidney Jesup reported that "the people on the frontiers of Texas and in New Mexico are prevented by the hostile Indians from cultivating their lands, the expense of maintaining the troops may be increased, but cannot be diminished."[143] Additionally, the federal government had to "protect the lines of emigration to New Mexico and Oregon . . . with mounted riflemen," and send weapons and troops to Oregon, where settlers were surrounded with Indian tribes."[144] One colonel of ordnance said that "although the supply of arms on hand may appear large, I am of opinion that it should be kept up and increased by manufacturing more annually than is requisite for ordinary consumption."[145] He meant government consumption, but ordinary consumption in general was changing. The years following the Mexican-American War were the beginning of a major civilian market for firearms. Pamela Haag locates the origins of this market in the second half of the nineteenth century in the strategies of arms makers who, unwilling to rely on fickle military demand, targeted a diverse consumer base with strategic sales and marketing.[146] Although Colt resented the government's decision not to renew his contract after the war, testimony from American soldiers who used his revolvers in Mexico became a major selling point.[147] Newspaper stories credited Colt's revolvers for US victory, which resulted from the "disparity of force between a thousand Americans armed with this deadly repeater, and three or four times the number of Mexicans aimed in the common manner."[148] One newspaper story quoted General Zachary Taylor's endorsement of Colt's revolvers as weapons that "may be relied upon under *all* circumstances," and noted that Taylor's opinion had been formed by men who "have performed feats of almost romantic daring and gallantry with them, during the war with Mexico."[149] One of Colt's first print advertisements from the early 1850s depicted scenes from the Mexican-American war.[150]

Colt's business took off in the years following the war, but it did so because of the precedent set by a half-century of government-sponsored industrial development. He borrowed machinery from Eli Whitney Jr., which the government had subsidized, to build a new factory in Hartford, Connecticut. Colt's later business ventures, along with those of Oliver Winchester, grew to be the most eminent private arms companies of the Civil War era and beyond. They benefited from private and public sales in a way that earlier manufacturers had not been able. The superiority of US arms manufacturing meant that Americans also found ready sales in Europe, a complete change from the days the United States imported Brown Besses and French Charlevilles. The federal armories were now capable of providing arms for troops within and beyond US borders, but the government could turn to the private sector if and when it needed.

It was becoming apparent that the United States was a major global player, and it produced and sold the industrial goods to prove it. The nineteenth century had opened with the United States anxious about and dependent on Britain. Halfway through the century, the opposite was true. There were still internal and external threats, to be sure, but those would never go away. If anything, those threats would continue to contribute to American industry.

Conclusion

THE UNITED STATES was the only country at the Crystal Palace Exhibition in London in 1851 that did not subsidize its participants. Despite its cavalier attitude toward the World's Fair, however, its 534 exhibitors won a greater share of prizes than any other nation.[1] Specimens of black cloth made from American fleece wool by S. Slater & Sons of Webster, Massachusetts, and blankets from Bay State Mills in Lawrence, Massachusetts, won prize medals.[2] Samuel Colt received great praise from Britain's Board of Ordnance for his revolving pistols. And the gold-medal carpet samples made by Erastus Bigelow's patented power loom caused the jury at the Exhibition to admit that "our American brethren have thus gained another step ahead of us."[3]

Times had changed. At the turn of the nineteenth century, the United States pirated technology from its former colonizer; fifty years later, Britain looked to emulate what it called the "American System of Manufactures." Likely originating in 1850s England, the label described the methods of interchangeability and mechanization that distinguished US production from the British factory system.[4] Three years after the Crystal Palace Exhibition, Britain's Board of Ordnance decided to model its new national armory on the Springfield Armory and to stock it with American-made machinery.[5] And Bigelow's award-winning carpets found homes in the White House and Congress. Less than twenty years earlier,

President Martin Van Buren's staff had still imported British carpeting for the White House, even though an act of 1826 stipulated that the White House should be furnished with as many articles of domestic manufacture as possible.[6]

US leaders understood well that independence and national security depended on Americans' ability to make, fix, and improve things. The individuals responsible for this system of manufacturing ranged from inventive mechanics in small New England towns and wealthy merchants in Boston to ordnance officials in Washington and consular agents in Lima, Peru. Aza Arnold's power loom improvements, combined with William Tudor's deployment of soft power in Peru, made possible an increase in cloth exports to South America, while Decius Wadsworth's relentless drive for weapon standardization and Simeon North's industriousness and network access made interchangeability in arms production a reality. And on a larger scale, the Madison Administration's decision to go to war with Britain spurred productivity with government contracts and market opportunities, while Secretary of State John Quincy Adams's negotiations with Spain for new territory provided capital for industrial development in Lowell, Massachusetts, and expanded US border patrol needs. These individuals were all part of the story of military and economic security in the early republic.

The relationships these men had to one another were important to the story. It mattered that Erastus Bigelow was not just an inventor from West Boylston, Massachusetts, but that he eventually worked for the Boston Associates in Lowell, who were connected to Secretary of State John Quincy Adams and US consul William Tudor. As an arms maker, Eli Whitney was an acquaintance of Secretary of State Oliver Wolcott and Colonel Decius Wadsworth, who in turn knew Nathan Starr, Simeon North, and Robert Johnson, all of whom traveled to Washington together to visit Secretary of War John C. Calhoun. These relationships mattered because they connected webs of manufacturers and mechanics to the administrative capacities of a burgeoning state. Federal officials declared war and negotiated treaties, to which the Hindsdill family in Bennington, Vermont, and Israel Thorndike in Salem, Massachusetts, responded—the Hindsdills by producing cloth for the military, Thorndike by investing merchant capital in new textile factories. These officials also maintained day-to-day relations with foreign countries, which made it possible for Nathan Appleton to ship cloth samples to his diplomatic associate in Peru and Roswell Lee to dispense armory surplus in Argentina. Federal officials' ability to manage geopolitical affairs required productivity and innovation in manufacturing, because without both, the United States could neither maintain an army nor serve as a valuable trade partner. The War and State Departments cultivated this productivity by issuing patents and contracts. When they recognized individuals like David Wilkinson, whose patented screw-making lathe helped mechanize both the arms and textile industries, and Simeon North, whose tenure as an arms contractor led to greater standardization in arms production and eventually mass production of textiles, they brought the nation closer to industrial superiority.

Federal officials engendered the public- and private-sector relationships that had created economic growth in the first half of the nineteenth century and that would make possible massive procurement for the Union Army during the Civil War. The mixed military economy that emerged in the 1860s had its roots in the industrial expansion of the first half of the nineteenth century.[7] The Mexican-American War was the ultimate test for military manufactures, and they passed it. By 1840, US output rivaled that of Great Britain and France, and by 1865 the Springfield Armory was the largest arsenal in the world.[8] The relationship between public and private changed in those years. Once the Ordnance Department had met its goals of standardization and the federal armories could fill both immediate and reserve needs, the regular letting of large government contracts to private arms manufacturers ceased.[9] The firms that had risen to prominence under federal patronage and paved the way in innovation and national service faded into obscurity.[10] They either ceased operations or, like Asa H. Waters and Company, downsized production and sold their guns to consumers elsewhere.[11] In general, these firms had difficulty adjusting to post-contract production. In their place, patent arms firms, most notably Winchester Repeating Arms Company and Colt's Patent Firearms Manufacturing, became the face of private arms manufacturing.[12]

While robust domestic output was a slower achievement in the textile industry than in the arms industry, the work that had gone into industrial development during the first half of the nineteenth century paid off militarily as the nation entered the second half. The United States became entirely self-sufficient in textile production by the time of the Civil War with the help of immigrant labor and increasing mechanization. For the first time in the nation's history, the federal government was able to provide its troops with uniforms made exclusively at home.[13] American textiles made their mark overseas as well. Sheetings that were made in Lowell and stamped with Lowell factory names sold well in China and South America, spurning imitators there. After the United States consolidated its Pacific coast landholdings, the Department of State could boast that US exports had moved into first rank in Chile's foreign trade and that the cotton textiles once imported almost exclusively from Britain were now supplied by the United States.[14] America's most noteworthy industry showed no signs of slowing. Amos Binney, who had procured cloth for American troops during the War of 1812, remarked in the 1840s that "the cotton manufacture is extending itself largely. New mills are erecting in every direction. A new city is projected on the rapids and falls of the Merrimack in Andover. Changes are so rapid and general in our neighborhood that from year to year one can hardly retain his recollection of familiar localities."[15] That new city was Lawrence, Massachusetts, whose blankets won prizes at the Crystal Palace Exhibit in London in 1851.

* * *

If Alexander Hamilton's 1791 *Report on Manufactures* had not set the nation on its industrial course when it called for "render[ing] the United States independent on foreign nations for military and essential supplies," it had at least predicted its outcome.[16] What Hamilton had said explicitly, the nation had done implicitly with a de facto partnership between federal officials and manufacturers. Government contracts, protective trade policies, immigration, and information sharing had generated war materiel and led to technological superiority. Shifting balances of power abroad created export opportunities for American wares, of which US diplomats and factory owners took advantage. Hamilton's vision for a nation that could bring to the international marketplace both agricultural and industrial produce was realized.

National security capitalism made this vision reality. Geopolitical considerations—a constellation of territory, rivalries, and resources—shaped the new nation's approach to political economy. Military and economic independence were of paramount importance following the Revolution. If the United States were to withstand European warfare, battles with Indians over land, and internal dissent, it would need to be able to clothe and arm its people. The United States had to prove its self-sufficiency to the outside world to avoid bullying. And so, as policy makers and manufacturers generated solutions to military and commercial conflicts, they created an industrial nation that by the 1850s was the envy of foreign onlookers. With industrial capabilities, the United States could take its place "among the powers of the earth."[17]

Industrial superiority was not a stated goal at the nation's founding; indeed, it would have seemed a lofty or impossible one. By the time of the Crystal Palace Exhibition in London in 1851, however, it was close to being achieved. Since before the nation's founding, New England's three major rivers had provided water power and transportation for entrepreneurs to build mills and send goods to market and for mechanics to tinker and develop new machinery. These waterways were just the start. At the same time as factory owners harnessed the water power by acquiring water privileges and building dams, federal officials exploited the water power to its fullest potential by employing the rivers' networks of producers, inventors, and capitalists. Members of the executive, from the president down to ordnance inspectors, toured these river valleys' factories, observing manufacturing processes, soliciting materials, and negotiating contracts. Their encouragement signaled that industrial improvement was a national imperative.

Of the three waterways, Blackstone's manufacturers received the least amount of direct federal support, in comparison to the large contracts that supported Connecticut Valley's arms makers or the claims payouts that were funneled into Merrimack. Still, Blackstone also benefited from patents, protective trade policies, and the occasional contract. As the century progressed, however, Merrimack Valley's industry eclipsed it; major capitalists in eastern Massachusetts had recourse to federal politics and diplomatic missions in a way that Blackstone producers did not. In the national story of industrialization, these disparities in access to policy makers became but a subplot as the United States began to surpass other nations in industrial capability. Job loss and worker alienation were the downsides of innovation, but the accumulation of all the improvements in manufacturing that were made in factories up and down these three rivers culminated in industrial strength that would bolster the nation's successes at the international bargaining table.[18]

The relationship between federal power and manufacturing capability in New England, of course, did not operate in a vacuum. Federal power connected with manufacturers wherever there was industry, just as manufacturers from New England were linked to merchants, consumers, and other factory owners in other regions. As the century progressed, the regions of the United States became more interconnected through markets, transportation, and manufacturing specialization. New England cemented its position as home to single-industry factory towns, supplying the nation with standardized textiles, shoes, guns, and clocks, while New York, Philadelphia, and New Jersey perfected the urban industrial model of small shops that produced highly specialized goods.[19] Delaware emerged as the capital of gunpowder production, while Pennsylvania became known for its iron industry. Local initiatives in the South employed slave labor and knowledge in the factory production of sugar, textiles, and other manufactured goods.[20] Each of these regions depended on the others for some combination of raw goods, capital, and consumption.

But it was in New England that the roots of the relationship between large-scale industrial business and the federal government were most firmly planted. Ties between government and the arms industry were strongest in New England, where the largest gun factories emerged. The largest textile factories, too, were in New England, and the connections between its textile capitalists and federal power predicted the relationship that would develop for the nation's first truly large business, the railroad.[21] Just as relationships between federal officials and producers and dependence on the government for contracts and markets characterized the development of the arms and textile industries, so they determined the emergence of the railroads. Some of the same capitalists who funded textile growth also dabbled in railroad development. When President John Quincy Adams in 1827 toured Thomas Handasyd Perkins's Granite Railway, which ran from Quincy to Charlestown in Massachusetts—the first commercial railroad in the United States—he remarked on Perkins's "ardent public spirit" in his diary.[22] Railroads, though, would not become "big business" until the 1860s, when some companies received the first federal corporate charters, and the government generated demand that did not exist among the general population.[23] War, too, created supply and contract opportunities for railroad companies, just as it had for antebellum manufacturing firms. Railroads would hasten connections among industries, consumers, and federal officials, and strengthen the links between territorial expansion, war, and security initiatives. By the end of the Civil War a new political economic order had emerged, held up by a powerful national state and the federal promotion of growth and large-scale enterprise.[24]

This economic order grew out of the river valleys of New England, where human capital and natural resources offered the nation the foundations of economic and military security. The region's manufacturers were receptive to a strong national state. Although the early twentieth century would witness a rewriting of the nation's history that minimized the role of the federal government in its foundation, this whitewashing did not reflect early republican reality.[25] Arms makers eagerly entered into federal contracts and adhered to stringent inspection standards; textile manufacturers took advantage of territorial expansion and federal commercial policies. They worked with federal officials to create a new nation that was not only politically independent but militarily independent as well. Its arms and textile technologies, which became the envy of European onlookers, were now capable of equipping a well-regulated army. The United States would never again be completely dependent on the Old World. Indeed, one 1842 survey characterized British industry as "once spectacular, now in distress," in contrast to its developing former colony.[26] As the United States developed advantages in manufacturing and, with it, political and military strength, the international marketplace became a favorable place for Americans to do business. Industry and the military were no longer anathema to the American experiment. The United States had made them work, in ways that still bedevil our experiment in government by, for, and of the people.

APPENDIX A
Terms Related to Textiles

Bombazet	Worsted cloth (cloth spun from fibers that have been combed to ensure that the fibers all run in the same direction) resembling bombazine (cloth made of silk warp and worsted weft in 2:1 twill weave), which could be twill or plain weave and was finished without glaze
Broadcloth	Carded wool in plain weave and fulled after weaving; woven on a wide loom and measured 54 to 63 inches in width
Brocade	Textiles with supplementary pattern wefts secured in the main ground weave; usually gold, silver, or silk, raised and enriched with flowers, foliage, or other ornaments
Calico	Cotton cloth of many grades and varieties, first made in India and later in Europe and America
Cambric	Fine white linen cloth in plain weave
Canvas	Unbleached, coarse cloth of hemp or flax
Cassimere	Medium-weight twilled woolen cloth of soft texture; plain varieties, competitors of broadcloth, were used for clothing and furnishing
Chambray	A type of gingham, plain in color and weave, often having a colored warp and white filling
Dimity	Any of a number of harness-loom patterned fabrics
Domestics	Cotton goods, shirting, and sheeting made in the United States

Source: Definitions are taken from Florence M. Montgomery, *Textiles in America, 1650–1870* (New York: W. W. Norton, 1984), 141–377.

Drab	Thick, stout, closely woven overcoating, which was heavy and expensive; also an undyed cloth of gray-beige color
Drab style	The color palette produced by dyeing cloth printed with different mordants to produce a variety of colors from yellow through buff, tan, and olive, to brown
Drill	Heavy linen cloth
Duck	Strong, thick linen cloth, finer and lighter than canvas
Flannel	Fabric made of woolen yarn, slightly twisted in the spinning to make the cloth soft
Fulling	Process of scouring and pressing whereby woolen goods were cleansed of grease and thickened into a compact finished material
Fustian	General term covering a large category of linen and cotton textiles, often made partly of wool
Gingham	Striped cloth woven with multiple-stranded warps and wefts and noted for toughness of texture; in India it was made of cotton and silk; in America it was made of pure cotton woven with dyed yarns
Jean	Linen/cotton, twilled cloth of the fustian group
Jersey	The finest wool taken out of other sorts of wool by combing it with a Jersey comb
Kersey	Cheap, coarse woolen cloth of twill weave
Linen	Cloth of many grades and weaves made from flax fibers
Madras	Goods exported from the Madras region of India. In the 1830s and 1840s, the term was associated with large, bright-colored handkerchiefs of silk warp and cotton woof (the crosswise yarns that fill the warp yarns in weaving)
Merino	Cloth woven from the wool of merino sheep

Muslin	Fine cotton textile first made in India
Muslinet	Fancy cotton dress material suggested for morning wraps, jackets, frocks, and high gowns
Nankeen	Cotton cloth of plain weave originally sold in Nanking, China, and made from a yellow variety of cotton
Naps	Heavy woolen material with raised surface twisted into little knots
Raven's duck	Coarse canvas or sailcloth
Sailcloth	Cloth used for sails and furniture for shipping
Sateen	Irregular twill weave in which the satin effect is produced by predominant weft threads
Satinet	Very slight, thin sort of satin, often inferior
Shoddy	Reclaimed wool, especially that obtained by pulling apart soft rags from worsted or from knitted or loosely woven woolens
Stuff	General term for worsted cloths
Ticking	Linen twill
Twill	A kind of weave producing a diagonal effect in the finished cloth
Velvet	A pile of fabric made of silk, wool, or cotton fibers
Velveteen	Cotton velvet
Warp	Set of yarns or other elements stretched in place on a loom before the weft is introduced during the weaving process; the longitudinal set in a finished fabric with two or more sets of elements
Weft	Threaded through the warp
Woolen	Cloth made of carded short-staple wool fibers
Worsted	Smooth cloth spun from long fibers that have been combed to make sure they all run parallel

APPENDIX B

Terms Related to Firearms

Ammunition	Shot, primers, powder, and cartridge cases used in charging firearms
Barrel	The metal tube of a gun whose purpose is to give the bullet velocity and direction, and to concentrate the gas that is generated by the explosion of the powder on the base of the bullet
Bayonet	Steel blade attached at the muzzle-end of a musket or rifle
Bore	Indication of caliber inside the barrel; determined by the number of perfectly round balls needed to weigh a pound
Breech	The action mechanism attached to the rear of the barrel
Brown Bess musket	Nickname for the British army's standard muzzle-loading .75-caliber smoothbore (unrifled barrel) flintlock long gun used in the eighteenth and early nineteenth centuries
Browning	Protection against rust; prevents light from reflecting off the firearm
Caliber	Diameter of the bore, expressed in decimal fractions of an inch
Cap	Container for an explosive charge; cover for the butt of a gun

Source: Definitions are taken from George Morgan Chinn Jr. and Bayless Evans Hardin, *Encyclopedia of American Hand Arms* (Huntington, WV: Standard Printing and Publishing, 1942), 29–32.

Carbine	A shorter-barreled version of a firearm, such as a rifle or musket
Cartridge	The container for an explosive charge that may or may not contain the bullet
Charleville musket (1763)	Standard French smoothbore flintlock long gun that had a slightly smaller barrel (.69-caliber) than the Brown Bess
Flintlock	A firearm that, with the spark made by the contact of flint against steel or iron, ignites the powder
Gauge	The diameter of the bore of a gun expressed in the number of balls of that diameter that are required to make a pound
Hammer	A device that strikes the firing pin or cartridge primer to detonate the powder
Handgun	Originally the term for hand cannon; any firearm that is easily carried
Lock	Refers to both the firing mechanism and to the part of the gun used to prevent firing by setting the safety
M1816	.69-caliber smoothbore flintlock musket produced by both the Springfield Armory and the Harpers Ferry Armory
Magazine	Part of a repeating firearm that holds cartridges to be fed into the gun
Matchlock	A firearm that requires a lighted match to ignite the powder
Model 1795 musket (US)	.69-caliber smoothbore flintlock musket produced by Springfield Armory—the first smoothbore modeled after the Charleville
Model 1816 musket (US)	Same as the M1816
Musket	A muzzle-loading firearm with a long barrel, fired from the shoulder

Muzzle	The forward end or mouth of a gun
Percussion cap	A small cylinder made of metal placed on the nipple or tube of a percussion gun holding fulminate (explosive salt made from organic acid). The sparks enter the barrel through a hole and fire the charge when struck
Pistol	A small firearm aimed and fired from one hand through a short barrel
Ramrod	A wood, iron, or steel rod for pushing the bullet into the muzzle-loader
Revolver	A hand arm, usually a pistol with a cylinder composed of several chambers arranged to revolve on an axis and discharge the shots in succession by the same lock
Rifle	A long-barreled firearm with groves cut into the bore walls to make a bullet spin and thus increase accuracy
Smoothbore	An unrifled bore
Springfield Model 1842	The last .69-caliber percussion musket manufactured by the United States

NOTES

Abbreviations

AAS	American Antiquarian Society
ASP	American State Papers
BLHC	Baker Library Historical Collections
CHS	Connecticut Historical Society
GPO	Government Printing Office
HSP	Historical Society of Pennsylvania
JRPP	Joel Roberts Poinsett Papers
MCHS	Middlesex County Historical Society
MHS	Massachusetts Historical Society
RIHS	Rhode Island Historical Society
RQMG	Records of the Office of the Quartermaster General
SA-LRM	Springfield Armory, Letters Received Miscellaneous
SA-LRO	Springfield Armory, Letters Received from Officials and Officers of the War and Treasury Departments
USDS-DC	US Department of State, Despatches [*sic*] from US consuls

Introduction

1. This book is about federal officials and manufacturing capitalists, all of whom were men. Women, of course, shaped political life, economic development, and territorial expansion. They worked in factories and occasionally owned stock in them. However important they were, this study is concerned with the individuals who pulled the levers of power in manufactories and political institutions.

2. Melvyn P. Leffler defines national security policy as "the decisions and actions deemed imperative to protect domestic core values from external threats." "National Security," *Journal of American History* 77, no. 1 (June 1990): 143. For industrial capacity as the foundation for military power, see Jonathan Kirshner, "Political Economy in Security Studies after the Cold War," *Review of International Political Economy* 5, no. 1 (Spring 1998): 66. For the importance of textiles in nations' security strategies, see Alan Collins, ed., *Contemporary Security Studies* (Oxford: Oxford University Press, 2013), 217.

3. In places where I refer in general to North American Native peoples,

rather than specific communities or nations, I use the terms "Indian" and "Native American" interchangeably. I recognize that these terms are imperfect, and they do not necessarily reflect the ways in which individuals chose to identify themselves.

4. For post-independence American anxieties about the United States' status in the world, see Kariann Akemi Yokota, *Unbecoming British: How Revolutionary America Became a Postcolonial Nation* (New York: Oxford University Press, 2011).

5. Paul A. C. Koistinen, *Beating Plowshares into Swords: The Political Economy of American Warfare, 1606–1865* (Lawrence: University Press of Kansas, 1996), 20.

6. While a government monopoly on war making was not a foregone conclusion, the executive branch early on established itself as the director of both public and private military undertaking and industries. Nicholas Parrillo, "De-Privatization of American Warfare: How the U.S. Government Used, Regulated, and Ultimately Abandoned Privateering in the Nineteenth Century," *Yale Journal of Law & the Humanities* 19 (Dec. 2007): 3–4. For the political economy of warfare in the nineteenth century, see Koistinen, *Beating Plowshares into Swords*; Mark R. Wilson, *The Business of Civil War: Military Mobilization and the State, 1861–1865* (Baltimore: Johns Hopkins University Press, 2006); William D. Adler and Andrew J. Polsky, "Building the New American Nation: Economic Development, Public Goods, and the Early U.S. Army," *Political Science Quarterly* 125, no. 1 (Spring 2010): 87–110; Andrew Fagal, "The Political Economy of War in the Early American Republic, 1774–1821" (PhD diss., Binghamton University, 2013); Andrew Fagal, "The Mills of Liberty: Foreign Capital, Government Contracts, and the Establishment of DuPont, 1790–1820," *Enterprise & Society* 19, no. 2 (June 2018): 309–351; Andrew Fagal, "American Arms Manufacturing and the Onset of the War of 1812," *New England Quarterly* 87, no. 3 (Sept. 2014): 526–537; and Andrew Fagal, "Terror Weapons in the Naval War of 1812," *New York History* 94, no. 3–4 (Summer/Fall 2013): 221–240. Max M. Edling has argued that the United States created a revenue system that allowed it to effectively mobilize resources for war and achieve dominance on the North American continent: *A Hercules in the Cradle: War, Money, and the American State, 1783–1867* (Chicago: University of Chicago Press, 2014). Political scientists have connected the War Department with innovation and economic development on the frontier. William D. Adler, "State Capacity and Bureaucratic Autonomy in the Early United States: The Case of the Army Corps of Topographical Engineers," *Studies in American Political Development* 26 (Oct. 2012): 110; Christopher Klyza, "The United States Army, Natural Resources, and Political Development in the Nineteenth Century," *Polity* 35, no. 1 (Autumn 2002): 1–28; and Stephen J. Rockwell, *Indian Affairs and the Administrative State in the Nineteenth Century*

(New York: Cambridge University Press, 2010). Military bureaucrats, more than individual pioneers, were responsible for frontier development. As they acted on behalf of a federal state that sought new commercial opportunities, a consolidation of power, and accurate knowledge of its territory, they helped build an expanding and prosperous republic. William H. Bergmann, *The American National State and the Early West* (New York: Cambridge University Press, 2012). Rachel St. John, however, cautions that "active" on the frontier did not necessarily mean "strong": "State Power in the West in the Early American Republic," *Journal of the Early Republic* 38, no. 1 (Spring 2018): 88.

7. This mattered especially because the textile industry, unlike other antebellum industries, such as railroads, was slow to attract investors, especially from abroad. Lance Edwin Davis, "Stock Ownership in the Early New England Textile Industry," *Business History Review* 32, no. 2 (Summer 1958): 204–222.

8. J. Thomas Scharf and Thomas Westcott, *History of Philadelphia, 1609–1884* (Philadelphia: L. H. Everts, 1884), 532.

9. Joyce Appleby, *The Relentless Revolution: A History of Capitalism* (New York: W. W. Norton, 2010); and Jürgen Kocka, *Capitalism: A Short History* (Princeton: Princeton University Press, 2016). For some of the most recent scholarship on capitalism in the United States, see Sven Beckert and Christine Desan, eds., *American Capitalism: New Histories* (New York: Columbia University Press, 2018).

10. For the limits of federalism, see Naomi R. Lamoreaux and John Joseph Wallis, "States, Not Nation: The Sources of Political and Economic Development in the Early United States," *Johns Hopkins Institute for Applied Economics, Global Health, and the Study of Business Enterprise, American Capitalism Working Papers*, no. 1 (Sept. 2015), 6, 20.

11. James Fallows, "The Military-Industrial Complex," *Foreign Policy* 133 (Nov.–Dec. 2002): 46. Priya Satia has recently located the origins of a "military-industrial complex" in eighteenth-century Britain: *Empire of Guns: The British State, the Industrial Revolution, and the Conscience of a Quaker Gun-Manufacturer* (New York: Penguin Random House, 2018).

12. Defense spending was 48 percent of federal expenditures in the 1960s. Appendix Table B-80, Economic Report of the President 2013, presentation by Robert C. Levine, Yale School of Management, January 28, 2014.

13. According to "Table Ea636-643—Federal Government Expenditure, by Major Function: 1789–1970," in Scott Sigmund Gartner, Michael R. Haines, Alan L. Olmstead, Richard Sutch, and Gavin Wright, *Historical Statistics of the United States*, ed. by Susan B. Carter (New York: Cambridge University Press, 2006), in 1810, federal defense spending constituted more than 55 percent of the budget: the army represented 24 percent, the navy 32 percent. In 1820, it was 48 percent of total expenditures: the army absorbed 28 percent, and the navy 20 percent. In 1830, the army represented over 30 percent of

government spending, the navy over 20 percent. During the War of 1812, defense spending spiked, totaling 80 percent of federal expenditures.

14. I define "industrial capitalism" as the prevalence of financial investment in mechanized production and the exchange of manufactured goods in the "free" market. This definition recognizes, however, that this wealth need not be private, and that "free" market exchange was determined by political decisions and social relationships. For the simultaneous development of nation-states and industrial capitalism, see, for example, Charles Tilly, *The Formation of National States in Western Europe* (Princeton: Princeton University Press, 1975).

15. Naomi Lamoreaux, "Rethinking the Transition to Capitalism in the Early American Northeast," *Journal of American History* 90, no. 2 (Sept. 2003): 437–461; and Christopher Clark, "Household Economy, Market Exchange and the Rise of Capitalism in the Connecticut Valley, 1800–1860," *Journal of Social History* 13, no. 2 (Dec. 1979): 169–189.

16. The role of the state in the economy has ebbed and flowed in the historiography of the early United States, and scholars coming from such fields as American Political Development (APD) and the "new history of capitalism" have weighed in. For the most recent theorizing about state power in the early republic, see the excellent series of essays in "Taking Stock of the State in Nineteenth-Century America," *Journal of the Early Republic* 38, no. 1 (Spring 2018): 61–118. A preoccupation with the links between capitalism and the state in early America goes back at least as far as the 1940s, to Oscar and Mary Handlin's pioneering study on the positive role of the Massachusetts state government on economic development. Oscar Handlin and Mary Flug Handlin, *Commonwealth: A Study of the Role of Government in the American Economy: Massachusetts, 1774–1861* (Cambridge: Belknap Press of Harvard University Press, 1969). For APD literature, see Stephen Skowronek, *Building a New American State: The Expansion of National Administrative Capacities, 1877–1920* (Cambridge: Cambridge University Press, 1982); Karen Orren and Stephen Skowronek, *The Search for American Political Development* (Cambridge: Cambridge University Press, 2004); Morton Keller, *America's Three Regimes: A New Political History* (New York: Oxford University Press, 2007); and Richard John, "Governmental Institutions as Agents of Change: Rethinking American Political Development in the Early Republic, 1787–1835," *Studies in American Political Development* 11 (Fall 1997): 347–380. See also William J. Novak, "The Myth of the Weak American State," *American Historical Review* 113, no. 3 (June 2008): 752–772. Other scholars, too, have noted the puzzling but stubborn notion of the United States as stateless. Desmond King and Robert C. Lieberman, "Ironies of State Building: A Comparative Perspective on the American State," *World Politics* 61, no. 3 (July 2009): 547–588. An activist state fell out of favor in the 1980s, when a body of scholars of the early republic described the transition to capitalism and the rise of industry in New England as

a story of Jeffersonian democracy ushering in individual freedoms, weak government, and ambitious entrepreneurship. See, for example, Joyce Appleby, *Capitalism and a New Social Order: The Republican Vision of the 1790s* (New York: New York University Press, 1984), and Gordon Wood, *The Radicalism of the American Revolution* (New York: Knopf, 1992). More recently, the "new history of capitalism" proceeds from the assumption that the state helped govern capitalist relations. Seth Rockman, "What Makes the History of Capitalism Newsworthy?" *Journal of the Early Republic* 34, no. 3 (Fall 2014): 443; Aaron Carico and Dara Orenstein, "Editors' Introduction: The Fictions of Finance," *Radical History Review* 118 (Winter 2014): 3–13; Joshua D. Rothman, *Flush Times and Fever Dreams: A Story of Capitalism and Slavery in the Age of Jackson* (Atlanta: University of Georgia Press, 2012); Kenneth Lipartito, "Connecting the Cultural and the Material in Business History," *Enterprise and Society* 14, no. 4 (Dec. 2013): 686–704; Michael Zakim and Gary J. Kornblith, eds., *Capitalism Takes Command: The Social Transformation of Nineteenth-Century America* (Chicago: University of Chicago Press, 2012); Sven Beckert, *Empire of Cotton: A Global History* (New York: Knopf, 2014); and Sven Beckert and Seth Rockman, *Slavery's Capitalism: A New History of American Economic Development* (Philadelphia: University of Pennsylvania Press, 2016). For recent work on the links between mercantile wealth and the development of federal power, see Gautham Rao, *National Duties: Customs Houses and the Making of the American State* (Chicago: University of Chicago Press, 2016). For the relationship among consumerism, the government, and capitalist development, see Joanna Cohen, *Luxurious Citizens: The Politics of Consumption in Nineteenth-Century America* (Philadelphia: University of Pennsylvania Press, 2017).

17. Historians have questioned whether the term "mercantilism" is even meaningful anymore. Philip J. Stern and Carl Wennerlind, eds. *Mercantilism Reimagined: Political Economy in Early Modern Britain and its Empire* (Oxford: Oxford University Press, 2013). For some of the debates about the eighteenth-century prevalence of mercantilist policy, see "Forum: Rethinking Mercantilism," *William and Mary Quarterly* 69, no. 1 (Jan. 2012): 3–70. Steve Pincus, for example, challenges the idea of a mercantilist consensus that governed British imperial policy: "Rethinking Mercantilism: Political Economy, the British Empire, and the Atlantic World in the Seventeenth and Eighteenth Centuries," *William and Mary Quarterly* 69, no. 1 (Jan. 2012): 3–34. Cathy Matson, Margaret Newell, Susan Amussen, and Trevor Burard, on the other hand, find some persistent meaning in the term, even as they highlight the complexity of ideological positions and policies that operated throughout the British Empire: "Imperial Political Economy: An Ideological Debate and Shifting Practices," *William and Mary Quarterly* 69, no. 1 (Jan. 2012): 35–40. Jonathan Barth suggests that there were multiple mercantilisms, just as there are multiple

capitalisms: "Reconstructing Mercantilism: Consensus and Conflict in British Imperial Economy in the Seventeenth and Eighteenth Centuries," *William and Mary Quarterly* 73, no. 2 (Apr. 2016): 261.

18. Mark R. Wilson, "Law and the American State, from the Revolution to the Civil War: Institutional Growth and Structural Change," in Michael Grossberg and Christopher Tomlins, eds., *The Cambridge History of Law in America*, vol. 2 (Cambridge: Cambridge University Press, 2008), 35.

19. The United States' trajectory toward sovereignty in some ways fit the European model that Charles Tilly laid out, in which states' four responsibilities are war making, state making, protection, and extraction: "'War Making and State Making as Organized Crime,'" in Theda Skocpol, Peter Evans, and Dietrich Rueschemeyer, eds., *Bringing the State Back In* (Cambridge: Cambridge University Press, 1985), 181. For the relationship between economic interest and liberal internationalism, see Glenda Sluga, "'Who Hold the Balance of the World?' Bankers at the Congress of Vienna, and in International History," *American Historical Review* 122, no. 1 (Dec. 2017): 1403–1430. For the contingency of the relationship between capitalism and the sovereign state, see Jonathan Levy, "Appreciating Assets: New Directions in the History of Political Economy," *American Historical Review* 122, no. 1 (Dec. 2017): 1490–1499.

20. John E. Crowley argues that before, during, and after the war, Americans preferred to maintain their privileged relationship with British or European metropole, rather than achieve manufacturing self-sufficiency or free trade: *The Privileges of Independence: Neomercantilism and the American Revolution* (Baltimore: Johns Hopkins University Press, 1993). Margaret Newell, on the other hand, characterizes New Englanders as economic liberals whose self-interested commercialism helped the colonists achieve independence: *From Dependency to Independence: Economic Revolution in Colonial New England* (Ithaca: Cornell University Press, 1998).

21. For an excellent overview of the scholarly debates surrounding the classification of the early American economy as capitalist, see Cathy Matson, ed., *The Economy of Early America: Historical Perspectives and New Directions* (University Park: Pennsylvania State University Press, 2006), 33–38, 57–58.

22. For economic thought following the Revolution, see Cathy D. Matson, "Capitalizing Hope: Economic Thought and the Early National Economy," *Journal of the Early Republic* 16, no. 2 (Summer 1996): 273–291.

23. Despite historians' disagreement over the usefulness of the term "mercantilism," Margaret Newell's definition seems to accurately characterize some of the commercial actions taken by both European and American policy makers. Newell defines it as "a loose group of policies that aimed to increase the nation's exports and to replace imports with the produce of domestic industry, offered ways to achieve a positive balance of trade" in "Putting the 'Political' Back in Political Economy (This Is Not Your Parents' Mercantilism)," *William and Mary*

Quarterly 69, no. 1 (Jan. 2012): 59. According to John E. Crowley, Americans were actually more in favor of state regulation than their British counterparts. Crowley, *The Privileges of Independence*.

24. William Appleman Williams, "The Age of Mercantilism: An Interpretation of the American Political Economy, 1763 to 1828," *William and Mary Quarterly* 15, no. 4 (Oct. 1958): 426, 433–434.

25. Max Weber, *Economy and Society: An Outline of Interpretative Sociology*, vol. 1 (Berkeley: University of California Press, 1978), 283.

26. Alexander Hamilton, Federalist no. 70, *The Federalist Papers*, ed. Clinton Rossiter (New York: New American Library, 1961), 471.

27. Edward Stanwood, *American Tariff Controversies in the Nineteenth Century*, vol. 1 (Boston: Houghton Mifflin, 1903).

28. This had ramifications for US military conflicts. Samuel J. Watson's recent work reveals a professionalized officer corps that carried out violence beyond the purview of civilian decision making: *Jackson's Sword: The Army Officer Corps on the American Frontier, 1810–1821* (Lawrence: University Press of Kansas, 2012).

29. For the importance of early foreign relations for the development of American nationhood, see Eliga Gould, *Among the Powers of the Earth: The American Revolution and the Making of a New World Empire* (Cambridge: Harvard University Press, 2012). For a reconsideration of the meaning of isolationism, see Marie-Jeanne Rossignol, "Early Isolationism Revisited: Neutrality and Beyond in the 1790s," *Journal of American Studies* 29, no. 2 (Aug. 1995): 215–227; and Rosemarie Zagarri, "The Significance of the 'Global Turn' for the Early American Republic: Globalization in the Age of Nation-Building," *Journal of the Early Republic* 31 (Spring 2011): 1–37. Historians of the early American republic continue to pursue the importance of global connections and frameworks for the era. See, for example, the excellent series of articles by Konstantin Dierks, Nancy Shoemaker, Emily Conroy-Krutz, Rachel Tamar Van, and Courtney Fullilove in "Forum: Globalizing the Early American Republic," *Diplomatic History* 42, no. 1 (Jan. 2018): 17–108.

30. Drew McCoy's indispensable work on Jeffersonian political economy, along with the more recent work of Brian Balogh, reveals the ways in which domestic and foreign policy were intertwined: the federal government's management of international relations led to trade opportunities and territorial expansion in the first decades of the nineteenth century. Drew McCoy, *The Elusive Republic: Political Economy in Jeffersonian America* (Chapel Hill: University of North Carolina Press, 1980); and Brian Balogh, *A Government Out of Sight: The Mystery of National Authority in Nineteenth Century America* (New York: Cambridge University Press, 2009). Hannah Farber has demonstrated the importance of American public credit for diplomatic and commercial relations, highlighting the role of tribute payments by the US

government in maintaining commercial markets in the Mediterranean. Farber, "Millions for Credit: Peace with Algiers and the Establishment of America's Commercial Reputation Overseas, 1795–96," *Journal of the Early Republic* 34, no. 2 (Summer 2014): 187–217. And Francis Cogliano argues that Thomas Jefferson achieved his "empire of liberty" through geopolitical means: *Emperor of Liberty: Thomas Jefferson's Foreign Policy* (New Haven: Yale University Press, 2014).

31. Caitlin A. Fitz, *Our Sister Republics: The United States in an Age of American Revolution* (New York: W. W. Norton, 2016); and James E. Lewis Jr., *The American Union and the Problem of Neighborhood: The United States and the Collapse of the Spanish Empire* (Chapel Hill: University of North Carolina Press, 1998).

32. Lawrence Delbert Cress argues that fears of centralized power led to the celebration of the militia. *Citizens in Arms: The Army and the Militia in American Society to the War of 1812* (Chapel Hill: University of North Carolina Press, 1982).

33. For white fears of Native Americans, see Robert M. Owens, *Red Dreams, White Nightmares: Pan-Indian Alliances in the Anglo-American Mind, 1763–1815* (Norman: University of Oklahoma Press, 2015). For fears of alliances between enslaved Americans and Spaniards, see Eric Herschthal, "Slaves, Spaniards, and Subversion in Early Louisiana: The Persistent Fears of Black Revolt and Spanish Collusion in Territorial Louisiana, 1803–1812," *Journal of the Early Republic* 36, no. 2 (Summer 2016): 283–311.

34. Max Edling, *A Revolution in Favor of Government: Origins of the U.S. Constitution and the Making of the American State* (New York: Oxford University Press, 2003), and "'So Immense a Power in the Affairs of War': Alexander Hamilton and the Restoration of Public Credit," *William and Mary Quarterly* 3rd series, 64, no. 2 (Apr. 2007): 287–326. Throughout the first half of the nineteenth century the US government created a bureaucratic network that was able to supply war materials for increasingly imperialistic military operations. Cynthia Ann Miller, "The United States Army Logistics Complex, 1818–1845: A Case Study of the Northern Frontier" (PhD diss., Syracuse University, 1991).

35. C. Vann Woodward, "The Age of Reinterpretation," *American Historical Review* 66, no. 1 (Oct. 1960): 3–4.

36. Fareed Zakaria, "The Myth of America's 'Free Security,'" *World Policy Journal* 14, no. 2 (Summer 1997): 35–43.

37. Richard H. Kohn argues that during the early republic the military exerted a tremendous influence on American domestic life. *Eagle and Sword: The Federalists and the Creation of the Military Establishment in America, 1783–1802* (New York: Free Press, 1975), xiii.

38. Historians disagree over the usage of "frontier" versus "borderlands."

Although the new social historians spotlighted the multiplicity of agents and motives operating in various and ambiguous frontiers, the term "frontier" has more commonly been associated with Anglo-American dominance and colonial binaries, while "borderlands" often signifies more fluid zones of interaction. Andrew R. L. Cayton and Fredrika J. Teute, eds., *Contact Points: American Frontiers from the Mohawk Valley to the Mississippi, 1750–1830* (Chapel Hill: University of North Carolina Press, 1998); Pekka Hämäläinen and Samuel Truett, "On Borderlands," *Journal of American History* 98, no. 2 (Sept. 2011): 343–344; and Jeremy Adelman and Stephen Aron, "From Borderlands to Borders: Empires, Nation States, and the Peoples in Between in North American History," *American Historical Review* 104, no. 3 (June 1999): 815–816. David Silverman has recently and compellingly defined "frontier" as a "zone of contact in which indigenous people exercised significant and sometimes even disproportionate power and the outcome was uncertain and contested." David Silverman, *Thundersticks: Firearms and the Violent Transformation of Native America* (Cambridge: Belknap Press of Harvard University Press, 2016), 19. In general, all of these terms reflect larger epistemological shortcomings. Andrew Cayton, "Not the Fragments But the Whole," *William and Mary Quarterly* 69, no. 3 (July 2012): 514. It is not my intention to enter into the theoretical debates surrounding these terms, but for the purposes of linking contested territory and conflicts between Europeans, white Americans, and Native Americans to manufacturing, I consider "frontier" in the rather limited sense of a sparsely settled backcountry characterized by episodes of violent conflict, which required military and material support.

39. For "Manifest Destiny" as an American ideology from settlement to the twentieth century, see Anders Stephanson, *Manifest Destiny: American Expansionism and the Empire of Right* (New York: Hill and Wang, 1995).

40. For the military side of war preparations, see J. P. Clark, *Preparing for War: The Emergence of the Modern U.S. Army, 1815–1917* (Cambridge: Harvard University Press, 2017).

41. For the debate about the experiences of capitalism in the early republic, see, for example, Lamoreaux, "Rethinking the Transition to Capitalism"; Paul A. Gilje, "The Rise of Capitalism in the Early Republic," *Journal of the Early Republic* 16, no. 2 (Summer 1996): 159–181; Matson, "Capitalizing Hope"; Winifred Barr Rothenberg, *From Market-Places to a Market Economy: The Transformation of Rural Massachusetts, 1750–1850* (Chicago: University of Chicago Press, 1992); and Clark, "Household Economy."

42. The classification of two different models or regions of industry in New England, with a third region in the Mid-Atlantic, can be traced back to James Montgomery, a manufacturer from Scotland who worked as a superintendent of a mill in Maine. He distinguished between an eastern district, which became known as the Waltham-Lowell System, a middle district, which became known

as the Rhode Island System, and a southern district outside of New England. Machinery, capitalization, and factory organization differentiated the three models. Later historians perpetuated the apparent differences between the Waltham-Lowell System, in which factories were financed by stockholders, managed by factory agents, and integrated all production processes under one roof, and the Rhode Island System, in which factories were financed and managed by small partnerships or sole proprietorships and which "put out" manufacturing processes, such as weaving, to individual households. James Montgomery, *A Practical Detail of the Cotton Manufacture of the United States of America* (Glasgow: John Niven, 1840).

43. Timothy Pitkin, *A Statistical View of the Commerce of the United States of America: Including also an Account of Banks, Manufactures and Internal Trade and Improvements* (New Haven: Durrie & Peck: 1835), 526.

44. For the destabilization of New England as the birthplace of industry, see François Weil, "Capitalism and Industrialization in New England, 1815–1845," *Journal of American History* 84, no. 4 (March 1998): 1336–1354. See also Thomas Childs Cochran, *Frontiers of Change: Early Industrialism in America* (New York: Oxford University Press, 1981). Cochran deemphasizes New England and textile production and argues that Delaware was an important center of industrialization.

45. What is missing from most studies of American industrialization is this geopolitical context. They have tended to focus on the social implications wrought by economic transformations for the worker, entrepreneur, firm, and community. Examples of studies of individual merchant-industrialists include Frances W. Gregory, *Nathan Appleton: Merchant and Entrepreneur, 1779–1861* (Charlottesville: University of Virginia Press, 1975); Chaim M. Rosenberg, *Life and Times of Francis Cabot Lowell, 1775–1817* (Lanham, MD: Lexington Books, 2011); Carl Seaburg and Stanley Paterson, *Merchant Prince of Boston: Colonel T. H. Perkins, 1764–1854* (Cambridge: Harvard University Press, 1971); and Barbara M. Tucker, *Samuel Slater and the Origins of the American Textile Industry, 1790–1860* (Ithaca: Cornell University Press, 1984). Works on the social and cultural changes that accompanied a rise in the factory system include Mary H. Blewett, *Men, Women, and Work: Class, Gender, and Protest in the New England Shoe Industry, 1780–1910* (Urbana: University of Illinois Press, 1988); Alan Dawley, *Class and Community: The Industrial Revolution in Lynn* (Cambridge: Harvard University Press, 1976); Thomas Dublin, *Transforming Women's Work: New England Lives in the Industrial Revolution* (Ithaca: Cornell University Press, 1994); Gary Kulik, "Pawtucket Village and the Strike of 1824: The Origins of Class Conflict in Rhode Island," *Radical History Review* 17 (1978): 5–37; and David A. Zonderman, *Aspirations and Anxieties: New England Workers and the Mechanized Factory System, 1815–1850* (New York: Oxford University Press, 1992). For business and technology, see Naomi R.

Lamoreaux, "Entrepreneurship, Business Organization, and Economic Concentration," in Stanley L. Engerman and Robert E Gallman, eds. *The Cambridge Economic History of the United States,* vol. 2. (Cambridge: Cambridge University Press, 2000) 403–434; James L. Conrad Jr., "Entrepreneurial Objectives, Organizational Design, Technology and Cotton Manufacturing of Almy and Brown 1787–1797," *Business and Economic History* 2nd series, 13 (1984): 7–19; and David Hounshell, *From the American System to Mass Production, 1800–1932: The Development of Manufacturing Technology in the United States*, vol. 4 (Baltimore: Johns Hopkins University Press, 1985). For factories as sites of business modernization, the iconic example is Alfred D. Chandler, *The Visible Hand: The Managerial Revolution in American Business* (Cambridge: Belknap Press of Harvard University Press, 1977).

46. Histories of early American textile manufacturing by William Bagnall, Victor Clark, and Caroline Ware have provided students of industrialization with comprehensive analyses of the nature of textile production in New England, while studies of the arms industry in Connecticut, Massachusetts, and at the federal armory in Springfield by Felicia Johnson Deyrup, James B. Whisker, and Kevin Spiker offer exhaustive details about arms manufacturing in New England. William R. Bagnall, *The Textile Industries of the United States* (Cambridge, MA: Riverside, 1893); Victor Selden Clark, *History of Manufactures in the United States: 1607–1860*, vol. 2 (Pittsburgh: Carnegie Institution, 1916); Caroline Ware, *The Early New England Cotton Manufacture, a Study in Industrial Beginnings* (Boston: Houghton Mifflin, 1931); Felicia Johnson Deyrup, *Arms Makers of the Connecticut Valley: A Regional Study of the Economic Development of the Small Arms Industry, 1798–1870* (Northampton, MA: Smith College Studies, 1948); James B. Whisker, *The United States Armory at Springfield, 1795–1865* (Lewiston, NY: E. Mellen Press, 1997); and James B. Whisker and Kevin Spiker, *The Arms Makers of Massachusetts, 1610–1900* (Palo Alto, CA: Academia Press, 2012). Merritt Roe Smith's now forty-year-old work on technological change at the federal armory at Harpers Ferry, Virginia, is still the standard bearer of scholarship on the development of the arms and machine tool industry. It provides a "microcosmic view of the industrial revolution" by focusing on how local customs and relationships shaped industrial change. My work, on the other hand, connects industrialization in the east with conflict and concerns over battlefield efficiency on the frontier. Merritt Roe Smith, *Harpers Ferry Armory and the New Technology: The Challenge of Change* (Ithaca: Cornell University Press, 1977).

47. Philip Scranton, *Proprietary Capitalism: The Textile Manufacture at Philadelphia, 1800–1885* (London: Cambridge University Press, 1983).

48. Anthony F. C. Wallace, *The Growth of an American Village in the Early Industrial Revolution* (New York: Knopf, 1978).

49. Historian François Weil argues that historians have mischaracterized

American industrialization by using Boston-centered New England generalizations that do not apply to the rest of the nation, or even to all of New England: "Capitalism and Industrialization in New England, 1815–1845." For an in-depth study of a mill town that differed markedly from the Lowell paradigm, see Wallace, *Growth of an American Village*.

50. Smith, *Harpers Ferry Armory and the New Technology*.

51. S. D. Smith, "The Market for Manufactures in the Thirteen Continental Colonies, 1698–1776," *Economic History Review*, New Series, 51, no. 4 (Nov. 1998): 679; and Peter Maw, "Yorkshire and Lancashire Ascendant: England's Textile Exports to New York and Philadelphia, 1750–1805," *Economic History Review* 63, no. 3 (Aug. 2010): 735.

52. John Maynard Keynes, *General Theory of Employment, Interest and Money* (London: Atlantic Books, 1936).

53. Probate inventories reveal greater percentages of guns than Michael Bellesiles's discredited study of gun ownership in early America showed, but few arms manufacturers had an easy time profiting from the civilian market. Michael A. Bellesiles, "The Origins of Gun Culture in the United States, 1760–1865," *Journal of American History* 83, no. 2 (Sept. 1996): 425–455; Michael A. Bellesiles, *Arming America: The Origins of a National Gun Culture* (New York: Knopf, 2000); and Mark D. Groover, "The Gibbs Farmstead: Household Archaeology in an Internal Periphery," *International Journal of Historical Archaeology* 9, no. 4 (Dec. 2005): 229–289. Maine-born arms manufacturer John Hancock Hall, for example, went into debt financing his business in the 1810s and subsequently turned to government contracting. R. T. Huntington, *Hall's Breechloaders* (York, PA: George Shumway, 1972), 9. None of this is to say that Americans did not buy guns. From the earliest days of colonization, European-Americans and Native Americans purchased and traded guns to hunt and fight with. Additionally, we cannot discount the link throughout American history between gun violence and white power, illustrated in Roxanne Dunbar-Ortiz's recent *Loaded: A Disarming History of the Second Amendment* (San Francisco: City Lights, 2018). Rather, I am arguing that there was not enough "natural" demand to entice the expansion of production. Carl P. Russell, *Guns on the Early Frontiers: A History of Firearms from Colonial Times through the Years of the Western Fur Trade* (Berkeley: University of California Press, 1957); and Merwyn Carey, *American Firearms Makers* (New York, Thomas Y. Crowell, 1953). For the "gun frontier," see Silverman, *Thundersticks*, 18.

54. In some respects, this would remain the case until the Civil War, as the United States' biggest export was cotton, and England its biggest consumer. Regardless of the importance of its cotton exports, the United States transitioned out of its status as solely an agricultural supplier.

Chapter 1. "Our Naked Troops"

1. Wayne Bodle, *The Valley Forge Winter: Civilians and Soldiers in War* (University Park: Pennsylvania State University Press, 2002). For suffering related to wartime atrocities, see Holger Hock, *Scars of Independence: America's Violent Birth* (New York: Crown, 2017). For Americans' memory of the military aspects of the revolutionary experience, see Sarah J. Purcell, *Sealed with Blood: War, Sacrifice, and Memory in Revolutionary America* (Philadelphia: University of Pennsylvania Press, 2002).

2. Pennsylvania Council of Safety, December 3, 1776, "Resolved, that Messieurs Christopher Marshall, Thomas Cuthbett, John Mifflin, Paul Fooks, John Lownes be appointed to go round the several wards of this city, and the Southern and Northern Liberties to collect all the old great coats, coats, surtouts, jackets and breeches from the inhabitants which they can spare" (Philadelphia: John Dunlap, 1776).

3. George Washington to Joseph Reed, December 27, 1778, in John C. Fitzpatrick, ed., *The Writings of George Washington from the Original Manuscript Sources, 1745–1799*, vol. 13 (Washington: US Government Printing Office [hereafter GPO], 1934), 457.

4. George Washington to Congress, September 5, 1778, in Fitzpatrick, ed., *The Writings of George Washington*, vol. 21, 403.

5. George Washington to Otis and Andrews, September 14, 1778, in Fitzpatrick, ed., *The Writings of George Washington*, vol. 21, 451–452.

6. George Washington to Benjamin Lincoln, January 9, 1781, in Fitzpatrick, ed., *The Writings of George Washington*, vol. 21, 74.

7. George Washington to Israel Shreve, January 9, 1781, in Fitzpatrick, ed., *The Writings of George Washington*, vol. 21, 76.

8. This cloth came from a woolen manufactory in Hartford, Connecticut, where Washington "perceive[d] a manufactory of them is established." George Washington to Henry Knox, January 29, 1789, in Frances Little, *Early American Textiles* (New York: Century, 1931), 104–105.

9. Philip Harling and Peter Mandler, "From 'Fiscal-military' State to Laissez-faire State, 1760–1850," *Journal of British Studies* 32, no. 1 (Jan. 1993): 44–70; and Patrick O'Brien, "Fiscal Exceptionalism: Great Britain and its European Rivals: From Civil War to Triumph at Trafalgar and Waterloo," in Donald Winch and Patrick O'Brien, eds., *The Political Economy of British Historical Experience: 1688–1914* (Oxford: Oxford University Press, 2002).

10. Armstrong Starkey, "Paoli to Stony Point: Military Ethics and Weaponry during the American Revolution," *Journal of Military History* 58, no. 1 (Jan. 1994): 11.

11. Sandra L. Powers, "Studying the Art of War: Military Books Known to American Officers and Their French Counterparts during the Second Half of

the Eighteenth Century," *Journal of Military History* 70, no. 3 (July 2006): 800, 788.

12. Stephen Conway, "The British Army, 'Military Europe,' and the American War of Independence," *William and Mary Quarterly* 67, no. 1 (Jan. 2010): 70–71.

13. Jessica Roney, "Government without Arms; Arms without Government: Philadelphia's Colonial Military Tradition" in Peter Onuf et al., eds. *Making Democracy: Violence and the American Founding* (Charlottesville: University of Virginia Press, 2014).

14. Nathan R. Kozushkanich, "Pennsylvania, the Militia, and the Second Amendment," *Pennsylvania Magazine of History and Biography* 133, no. 2 (Apr. 2009): 119.

15. Robert Middlekauff, "Why Men Fought in the American Revolution," *Huntington Library Quarterly* 43, no. 2 (Spring 1980): 141–144.

16. D. J. Smith, "Army Clothing Contractors and the Textile Industries in the 18th Century," *Textile History* 14, no. 2 (Jan. 1983): 162.

17. Victor S. Clark, *History of Manufactures in the United States* (Washington: Carnegie Institute, 1929), 5–6.

18. Smith, "Army Clothing Contractors," 154–155; and Jack P. Greene and J. R. Pole, eds., *The Blackwell Encyclopedia of the American Revolution* (Cambridge: Blackwell, 1991), 149.

19. For the relationship between British military needs and gun production, see Priya Satia, *Empire of Guns: The British State, the Industrial Revolution, and the Conscience of a Quaker Gun-Manufacturer* (New York: Penguin Random House, 2018).

20. Clark, *History of Manufactures in the United States*, 5–6.

21. Powers, "Studying the Art of War," 795.

22. Horatio Gates to John Adams, April 23 and 27, 1776, in Bernhard Knollenberg, "The Correspondence of John Adams and Horatio Gates," *Proceedings of the Massachusetts Historical Society* [hereafter MHS] 67 (Oct. 1941): 140–141.

23. Rolla M. Tyron, *Household Manufactures in the United States, 1640–1860* (Chicago: University of Chicago Press, 1917).

24. Clark, *History of Manufactures in the United States*, 33–35, 10.

25. Carl Bridenbaugh, *Cities in Revolt: Urban Life in America, 1743–1776* (New York: Knopf, 1955), 268.

26. Clark, *History of Manufactures in the United States*, 34–35.

27. Margaret Newell, *From Dependency to Independence: Economic Revolution in Colonial New England* (Ithaca: Cornell University Press, 1998), 297.

28. Felix Reichmann, "Pennsylvania Rifle," *Pennsylvania Magazine History Biography* 69, no. 3 (Jan. 1945).

29. Neil L. York, "Pennsylvania Rifle: Revolutionary Weapon in a Conven-

tional War?" *Pennsylvania Magazine of History and Biography* 103, no. 3 (July 1979): 303–305.

30. John W. Wright, "The Rifle in the American Revolution," *American Historical Review* 29, no. 2 (Jan. 1924): 294.

31. Merrill Lindsay, *The New England Gun: The First Two Hundred Years* (New Haven: New Haven Colony Historical Society, 1975), 2–7.

32. Robert Martello, *Midnight Ride, Industrial Dawn: Paul Revere and the Growth of American Enterprise* (Baltimore: Johns Hopkins University Press, 2010); Donald E. Reynolds, "Ammunition Supply in Revolutionary Virginia," *Virginia Magazine of History and Biography* 73, no. 1 (Jan. 1965): 56; and Esther Forbes, *Paul Revere and the World He Lived In* (Boston: Houghton Mifflin Harcourt, 1999), 302–303.

33. Clark, *History of Manufactures in the United States,* 34; New York Provincial Congress, "Resolutions of the Provincial Congress, of the colony of New-York, for the encouragement of manufactories of gun powder, musket barrels, musket locks, and salt" (New York: John Holt, 1776).

34. Massachusetts General Court, House of Representatives, "A journal of the Honorable House of Representatives of the colony of the Massachusetts-Bay in New-England" (Watertown, MA: Benjamin Edes, 1775), 13–14; and New York Committee of Safety, Oct. 9, 1776, "Resolved, that the persons hereafter mentioned, be appointed to purchase at the cheapest rate, in their several counties, all the coarse woollen cloth, linsey-woolsey, blankets" (Fishkill, NY: Samuel Loudon, 1776).

35. Pennsylvania Council of Safety, "To the public" (Philadelphia: William and Thomas Bradford, 1776); and Pennsylvania Council of Safety, December 3, 1776.

36. Paul A. C. Koistinen, *Beating Plowshares into Swords: The Political Economy of American Warfare, 1606–1865* (Lawrence: University Press of Kansas, 1996), 21; and *Journals of the Continental Congress*, October 2, 1775.

37. John Adams diary 26, February 16–April 1776 [electronic edition]. Adams Family Papers: An Electronic Archive, MHS, http://www.masshist.org/digitaladams.

38. Elizabeth Cometti, "The Civil Servants of the Revolutionary Period," *Pennsylvania Magazine of History and Biography* 75, no. 2 (Apr. 1951): 160; Kurt Daniel Kortenhof, "Republican Ideology and Wartime Reality: Thomas Mifflin's Struggle as the First Quartermaster General of the Continental Army, 1775–1778," *Pennsylvania Magazine of History and Biography* 122, no. 3 (July 1998): 189; and Pennsylvania Board of War, "To the Public," May 2, 1777 (Philadelphia: John Dunlap, 1777).

39. American State Papers [hereafter ASP], "Claims for Losses Sustained by General Greene in Procuring Supplies for the Southern Army, in 1782," December 26, 1791, no. 23: 37.

40. David L. Salay argues that Congress effectively mobilized manufacturing capabilities in Pennsylvania. "Marching to War: The Production of Leather and Shoes in Revolutionary Pennsylvania," *Pennsylvania History: A Journal of Mid-Atlantic Studies* 60, no. 1 (Jan. 1993): 52.

41. "A report of the Board of War," August 13, 1776, Founding Families: Digital Editions of the Papers of the Winthrops and the Adamses, ed. C. James Taylor (Boston: MHS, 2018); Richard G. Leonard, "The Clothing Situation in the American Revolutionary Army" (MA thesis, Loyola University Chicago, 1948), 26–28.

42. Kenneth Schaffel, "The American Board of War, 1776–1781," *Military Affairs* 50, no. 4 (Oct. 1986): 186.

43. *Journals of the Continental Congress,* May 23, 1776.

44. Schaffel, "The American Board of War," 186–188.

45. Adams generally read theoretical economists for specific points of policy rather than for overarching philosophies. Manning J. Dauer, "The Political Economy of John Adams," *Political Science Quarterly* 56, no. 4 (Dec. 1941): 550.

46. Dauer, "The Political Economy of John Adams," 550; and Gerard Clarfield, "John Adams: The Marketplace, and American Foreign Policy," *New England Quarterly* 52, no. 3 (Sept. 1979): 347.

47. Dauer, "The Political Economy of John Adams," 550; and "From John Adams to Stephen Sayre, December 6, 1780," and "To John Adams from Stephen Sayre, January 10, 1781," *Founders Online,* National Archives, last modified February 1, 2018, http://founders.archives.gov/documents/Adams/06-11-02-0023. Original source: Gregg L. Lint, Richard Alan Ryerson, Anne Decker Cecere, Celeste Walker, Jennifer Shea, and C. James Taylor, eds., *The Adams Papers, Papers of John Adams,* vol. 11, January–September 1781 (Cambridge: Harvard University Press, 2003), 36–39.

48. Bill of Lading, July 25, 1796, Binder 2, Jay T. Last Collection of Maritime Prints and Ephemera, Series 1, The Huntington Library, San Marino, California. Before the War of 1812, federal officials sent samples of Russian duck to American manufacturers as a model, which will be discussed in chapter 3.

49. Greene and Pole, eds., *The Blackwell Encyclopedia,* 322.

50. Wright, "The Rifle in the American Revolution," 295.

51. Reichmann, "Pennsylvania Rifle," 12.

52. Daniel Morgan's Provisional Rifle Corps from Virginia, however, continued to be instrumental in the Battles of Saratoga and Cowpens.

53. Wright, "The Rifle in the American Revolution," 297.

54. York, "Pennsylvania Rifle," 308, 314; and Samuel Hodgson to John Harris, September 3, 1798, National Archives and Record Administration: Post-Revolutionary War Papers, Record Group 45.

55. George Washington to Jedidiah Huntington, November 19, 1778, and George Washington to George Measam, November 25, 1778, in Fitzpatrick, ed., *The Writings of George Washington,* vol. 13, 286, 330; and United States, In Congress, March 23, 1779, "Ordinance for regulating the cloathing department for the armies of the United States" (Philadelphia: David C. Claypoole, 1779).

56. *Journals of the Continental Congress*, September 25, 1782; and George Washington to Benjamin Lincoln, January 9, 1781, in Fitzpatrick, ed., *The Writings of George Washington,* vol. 21, 74.

57. For the role of the new federal government in balancing economic freedom and civic responsibility, see Joanna Cohen, *Luxurious Citizens: The Politics of Consumption in Nineteenth-Century America* (Philadelphia: University of Pennsylvania Press, 2017).

58. *Journals of the Continental Congress*, August 19, 1783.

59. *Journals of the Continental Congress*, January 2, 1782; and "Mr. Hartley's Proposed Article of Agreement, delivered by him to the American Commissioners for their consideration, May 21, 1783" in Francis Wharton, ed., *The revolutionary diplomatic correspondence of the United States*, vol. 6 (Washington: GPO, 1889).

60. General economic downturn, however, meant that British sales to the United States declined $1.3 million annually for the six years following the war as compared with those preceding it. Clark, *History of Manufactures in the United States,* 229.

61. For the economic implications of the political transition from confederation to federalism, see Cathy D. Matson and Peter S. Onuf, *A Union of Interests: Political and Economic Thought in Revolutionary America* (Lawrence: University Press of Kansas, 1990).

62. For federal power and taxation, see Max M. Edling, *A Revolution in Favor of Government: Origins of the U.S. Constitution and the Making of the American State* (New York: Oxford University Press, 2003). For debates over states' rights to create monetary policy, see Woody Holton, *Unruly Americans and the Origins of the Constitution* (New York: Hill and Wang, 2008). For federal power and taxation, see Edling, *A Revolution in Favor of Government.*

63. James Madison, "Vices of the Political System of the United States," April 1787, in Gaillard Hunt, ed., *The Writings of James Madison: 1783–1787*, vol. 2 (New York, 1901), 364.

64. Robin L. Einhorn, *American Taxation, American Slavery* (Chicago: University of Chicago Press, 2006); Charles Beard, *An Economic Interpretation of the Constitution of the United States* (New York, 1913); Gordon Wood, *The Creation of the American Republic, 1776–1787* (Chapel Hill: University of North Carolina Press, 1969); Holton, *Unruly Americans*; and Edling, *A Revolution in Favor of Government*. All of these things can be true, because as Eric Slauter reminds us, our understanding of the Constitution is mediated by

interpretations of James Madison's edited notes on the Constitutional Convention and the transcribed reports of state ratifying conventions. Eric Slauter, *The State as a Work of Art: The Cultural Origins of the Constitution* (Chicago: University of Chicago Press, 2009), 23. The Constitution was ultimately a product of myopic particularities, even as it seems to offer a farsighted prescription for national governance. Slauter, *The State as a Work of Art,* 21. For the delegates' preoccupation with security and diplomacy, see Robbie J. Totten, "Security, Two Diplomacies, and the Formation of the U.S. Constitution: Review, Interpretation, and New Directions for the Study of the Early American Period," *Diplomatic History* 36, no. 1 (Jan. 2012): 79.

65. Lawrence A. Peskin, *Manufacturing Revolution: The Intellectual Origins of Early American Industry* (Baltimore: Johns Hopkins University Press, 2003).

66. Slauter, *The State as a Work of Art,* 20.

67. See for example, "the petition of James Rumsey, relating to his invention of a power boat" and "petition of Michael Byrne for Congress's patronage of his invention which was reported on by the Committee of the Week," *Journals of the Continental Congress*, January 17 and 26, 1785.

68. Dotan Oliar, "The Constitutional Convention on IP: A New Reading," *UCLA Law Review* 57, no. 2 (Nov. 2009): 445–446. For issues surrounding copyright policy in the early republic, see Nora Slonimksy, "The Engine of Free Expression: The Political Development of Copyright in the Colonial British Atlantic and Early National United States" (PhD diss., CUNY Graduate Center, 2017).

69. Lawrence Delbert Cress, *Citizens in Arms: The Army and the Militia in American Society to the War of 1812* (Chapel Hill: University of North Carolina Press, 1982).

70. Gaillard Hunt, ed., *The Writings of James Madison,* vol. 3: 1787, *The Journal of the Constitutional Convention,* part 1 (New York: Putnam, 1902), 185.

71. Max Farrand, ed., *The Records of the Federal Convention of 1787*, vols. 1 and 2 (New Haven: Yale University Press, 1911).

72. Morton J. Frisch, "Executive Power and Republican Government—1787," *Presidential Studies Quarterly* 17, no. 2 (Spring 1987): 282.

Chapter 2. The Political Economy of Guns and Textiles

1. Tench Coxe, *A brief examination of Lord Sheffield's Observations on the commerce of the United States* (Philadelphia: M. Carey, 1791), 3, 42.

2. Jacques-Pierre Brissot de Warville, *New Travels in the United States of America, 1788,* trans. Mara Soceanu Vamos and Durand Echeverria (Cambridge: The Belknap Press of Harvard University Press, 1964); and Walter E. Minchinton, *Silas Deane and Lord Sheffield's "Observations on American Commerce"* (Coimbra, Portugal: UC Biblioteca Geral 1, 1980), 83.

3. John E. Crowley, *The Privileges of Independence: Neomercantilism and the American Revolution* (Baltimore: Johns Hopkins University Press, 1993), 67. See also Kariann Akemi Yokota, *Unbecoming British: How Revolutionary America Became a Postcolonial Nation* (New York: Oxford University Press, 2011).

4. Richard H. Kohn, *Eagle and Sword: The Federalists and the Creation of the Military Establishment in America, 1783–1802* (New York: Free Press, 1975), xi–xiii.

5. Timothy Pitkin, *A Statistical View of the Commerce of the United States of America: Including also an Account of Banks, Manufactures and Internal Trade and Improvements* (New Haven: Durrie & Peck, 1835), 474.

6. Don Higginbotham, "The Federalized Militia Debate: A Neglected Aspect of Second Amendment Scholarship," *William and Mary Quarterly* 55, no. 1 (Jan. 1998): 40. See also Saul Cornell, *A Well-Regulated Militia: The Founding Fathers and the Origins of Gun Control in America* (New York: Oxford University Press, 2008), and Michael Waldman, *The Second Amendment: A Biography* (New York: Simon & Schuster, 2015).

7. ASP, "Militia," Military Affairs 1, no. 21 (Dec. 11, 1794): 69.

8. For tariffs as a solution to debates over taxation, particularly between "free" and "slave" states, see Robin L. Einhorn, *American Taxation, American Slavery* (Chicago: University of Chicago Press, 2006).

9. US Congress, "An act for laying a duty on goods, wares, and merchandizes imported into the United States" (New York: Childs and Swaine, 1789).

10. US Congress, "Duties payable by law, on all goods, wares & merchandize, imported into the United States of America, after the last day of June, 1797" (Philadelphia, 1797).

11. ASP, "Organization of the Militia," Military Affairs 1, no. 2 (Jan. 21, 1790): 13.

12. US Congress, "An act for establishing an executive department, to be denominated the Department of Foreign Affairs" (New York: Childs and Swaine, 1789).

13. See, for example, Annals of Congress, "Petition of Christopher Collis of the city of New York praying that an exclusive privilege may be granted him in the benefits of an invention . . . for counting with the utmost precision the number of revolutions or vibrations of any wheel or other part of the mechanical engine or machine," August 4, 1789, 1st Congress, 1st Session, 701.

14. Victor S. Clark, *History of Manufactures in the United States* (Washington: Carnegie Institute, 1929), 48.

15. James Rumsey to Thomas Jefferson, June 6, 1789, in Julian P. Boyd, ed., *The Papers of Thomas Jefferson*, vol. 15 (Princeton: Princeton University Press, 1958).

16. Thomas Jefferson to James Madison, September 6, 1789, in Herbert

Sloan, "The Earth Belongs in Usufruct to the Living," in Peter Onuf, ed., *Jeffersonian Legacies* (Charlottesville: University of Virginia Press, 1993), 285.

17. For Jefferson's opposition to patents, see, for example, Thomas Jefferson to James Madison, July 31, 1788, in James Smith, ed., *The Republic of Letters: The Correspondence of James Madison and Thomas Jefferson*, vol. 1 (New York: W. W. Norton, 1995). For the issue of monopolies and intellectual processes, see Nora Slonimksy, "The Engine of Free Expression: The Political Development of Copyright in the Colonial British Atlantic and Early National United States" (PhD diss., CUNY Graduate Center, 2017).

18. Patent Act of 1790, Ch. 7, 1 Stat. 109 (April 10, 1790).

19. Economic historian Zorina B. Khan argues that the accessibility of the American patent system, far from creating an inefficient cycle of patent litigation, fostered the dissemination of knowledge and ideas that led to economic growth. *The Democratization of Invention: Patents and Copyrights in American Economic Development, 1790–1920* (New York: Cambridge University Press, 2005).

20. Zorina B. Khan, "Premium Inventions: Patents and Prizes as Incentive Mechanisms in Britain and the United States, 1750–1930," in *Understanding Long-Run Economic Growth: Geography, Institutions, and the Knowledge Economy* (Chicago: University of Chicago Press, 2008), 211.

21. Khan, "Premium Inventions," 210.

22. Barbara M. Tucker, *Samuel Slater and the Origins of the American Textile Industry, 1790–1860* (Ithaca: Cornell University Press, 1984), 21; David J. Jeremy, *Transatlantic Industrial Revolution: The Diffusion of Textile Technologies between Britain and America, 1790–1830s* (Cambridge: MIT Press, 1981), 79; and Khan, "Premium Inventions," 207.

23. "The Rumford Premium of the American Academy," *Journal of the Franklin Institute*, vol. 1, New Series, Jan. 1828.

24. Naturalization Act of 1790, Ch. 2, Stat. 2, March 1, 1790.

25. Doron Ben-Atar, *Trade Secrets: Intellectual Piracy and the Origins of American Industrial Power* (New Haven: Yale University Press, 2004), 104; and Jeremy, *Transatlantic Industrial Revolution*, 51.

26. Benjamin Rush, "Information to Europeans who are disposed to migrate to the United States. In a letter from a citizen of Pennsylvania, to his friend in Great Britain" (Philadelphia: Carey, Stewart, 1790), 7.

27. Rush, "Information to Europeans," 7.

28. David J. Jeremy, *Henry Wansey and His American Journal, 1794* (Philadelphia: American Philosophical Society, 1970), 40, 62–63.

29. Minchinton, *Silas Deane*, 86. For Americans' persistent ambivalence about consumption of British goods, see Joanna Cohen, *Luxurious Citizens: The Politics of Consumption in Nineteenth-Century America* (Philadelphia: University of Pennsylvania Press, 2017).

30. Gerard Clarfield, "John Adams: The Marketplace, and American Foreign Policy," *New England Quarterly* 52, no. 3 (Sept. 1979): 354.

31. Jeremy, *Henry Wansey and His American Journal*, 127.

32. Pitkin, *A Statistical View*, 474.

33. I understand the Industrial Revolution to be a gradual transition to making things with mechanized manufacturing processes, the concentration of labor and capital in factories, and the development of new technologies that altered humans' relationships to their surroundings. It was not inevitable but was the product of a set of business and political decisions that generated great social, economic, and technological changes.

34. According to John R. Nelson, Hamilton actually opposed domestic manufacturing; his Society for Establishing Useful Manufactures was more about rechanneling capital away from risky securities than it was about industry. John R. Nelson Jr., *Liberty and Property: Political Economy and Policymaking in the New Nation, 1789–1812* (Baltimore: Johns Hopkins University Press, 1987).

35. Alexander Hamilton, *Report on Manufactures*, December 5, 1791, in Harold C. Syrett, ed., *The Papers of Alexander Hamilton*, vol. 19 (New York: Columbia University Press, 1973).

36. William Appleman Williams, "The Age of Mercantilism: An Interpretation of the American Political Economy, 1763 to 1828," *William and Mary Quarterly* 15, no. 4 (Oct. 1958): 430. Scholars have debated the extent to which Hamilton's policies were intended to stimulate manufacturing. For an overview of this discussion, see Andrew Shankman, "'A New Thing on Earth': Alexander Hamilton, Pro-Manufacturing Republicans, and the Democratization of American Political Economy," *Journal of the Early Republic* 23, no. 3 (Autumn 2003): 326. John R. Nelson argues that Hamilton was not truly pro-manufacturing because he failed to endorse protective tariffs, while Drew McCoy posited that Hamilton's support of bounties and premiums demonstrated his support for industry. John R. Nelson, "Alexander Hamilton and American Manufacturing: A Reexamination," *Journal of American History* 65, no. 4 (Mar. 1979): 971–995; and Drew McCoy, *The Elusive Republic: Political Economy in Jeffersonian America* (Chapel Hill: University of North Carolina Press, 1980). Hamilton's plan for bounties did not receive Congressional support, but most of his tariff recommendations were implemented in the early 1790s.They were not highly protectionist because Hamilton did not want to destroy the tax base by decreasing imports. Douglas A. Irwin, "The Aftermath of Hamilton's 'Report on Manufactures,'" *Journal of Economic History* 64 (Sept. 2004): 800–821.

37. Hamilton believed the military should be under national authority. Alexander Hamilton, "Concerning the Militia," *The Federalist* no. 29, *Independent Journal*, January 9, 1788.

38. Recent work by Andrew Fagal, however, contradicts this notion, as evidenced by Jefferson's vision for a political economy driven by military

production. "The Political Economy of War in the Early American Republic, 1774–1821" (PhD diss., Binghamton University, 2013).

39. McCoy, *The Elusive Republic.*

40. Ken Alder, *Engineering the Revolution: Arms and Enlightenment in France, 1763–1815* (Chicago: University of Chicago Press, 2010), 3–4. Andrew Fagal's research suggests that the Jeffersonian political economy was indeed responsible for the development of US war-making capabilities. Fagal, "The Political Economy of War."

41. McCoy, *The Elusive Republic,* 211.

42. McCoy, *The Elusive Republic,* 65.

43. Robert J. Allison, *Crescent Obscured: The United States and the Muslim World, 1776–1815* (Chicago: University of Chicago Press, 1995). See chapter one for an overview of American policy toward the Muslim world following American independence.

44. McCoy, *The Elusive Republic,* 174.

45. Sean Wilentz argues that mechanics developed a unique social and political ideology, fusing trade pride, Revolutionary ideology, and egalitarianism into what he calls "artisan republicanism." This ideology grew out of, and encouraged, common artisan political and social behaviors. Sean Wilentz, *Chants Democratic; New York City and the Rise of the American Working Class 1788–1850* (New York: Oxford University Press, 1984), 13. See also Howard B. Rock, Paul A. Gilje, and Robert Asher, eds., *American Artisans: Crafting Social Identity, 1750–1850* (Baltimore: Johns Hopkins University Press, 1995); Howard B. Rock, *Artisans of the New Republic: The Tradesmen of New York City in the Age of Jefferson* (New York: New York University Press, 1979); and Lawrence A. Peskin, "From Protection to Encouragement: Manufacturing and Mercantilism in New York City's Public Sphere, 1783–1795," *Journal of the Early Republic* 18, no. 4 (Winter 1998): 590.

46. Peskin, "From Protection to Encouragement," 594–614.

47. Shankman, "A New Thing on Earth," 340–341. Jeffersonians, too, came to recognize that social equality and capitalism were incompatible. Andrew Shankman, *Crucible of American Democracy: The Struggle to Fuse Egalitarianism & Capitalism in Jeffersonian Pennsylvania* (Lawrence: The University Press of Kansas, 2004).

48. ASP, "Increase of Duties," no. 65 (April 1794): 276–277. The federal tariff did much more to protect certain trades than the New York state tariff that it replaced. Peskin, "From Protection to Encouragement," 604.

49. ASP, "Drawback on Cordage," Finance 1, no. 48 (Feb. 7, 1793): 202.

50. ASP, "Loan to John F. Amelung," Finance: 1, no. 14 (June 2, 1790): 62.

51. Annals of Congress, "Encouragement to Manufacturers," January 9, 1797, House of Representatives, 4th Congress, 2nd Session, 1825–1826.

52. Indeed, the first congressional argument for the unconstitutionality of

tariffs to promote manufacturing interests came from Massachusetts Representative Ezekial Whitman, which illustrates the diversity of commercial interests even within a state that would come to develop substantial industry. Daniel Peart, "Looking beyond Parties and Elections: The Making of United States Tariff Policy during the Early 1820s," *Journal of the Early Republic* 33, no. 1 (Spring 2013): 106n27.

53. "Table Ee533-550—Exports, by country of destination: 1790–2001," in Scott Sigmund Gartner, Michael R. Haines, Alan L. Olmstead, Richard Sutch, and Gavin Wright, *Historical Statistics of the United States*, ed. Susan B. Carter (New York: Cambridge University Press, 2006).

54. Stanley Elkins and Eric McKitrick, *Age of Federalism* (New York: Oxford University Press, 1993), 375–389.

55. For negotiations and controversies surrounding the Jay Treaty, see Elkins and McKitrick, *Age of Federalism,* chapter nine.

56. McCoy, *The Elusive Republic,* 175.

57. Economists have debated the extent to which antebellum tariffs were responsible for the competitiveness of American textiles. C. Knick Harley contended that tariffs were necessary, while Peter Temin and Douglas A. Irwin argued the opposite. C. Knick Harley, "International Competitiveness of the Antebellum American Cotton Textile Industry," *Journal of Economic History* 52, no. 3 (Sept. 1992): 559–584; and Douglas A. Irwin and Peter Temin, "The Antebellum Tariff on Cotton Textiles Revisited," *Journal of Economic History* 61, no. 3 (Sept. 2001): 777–798.

58. Most respondents in the 1820 manufacturing census reported that their factories commenced operation after 1807. US Department of State, *Records of the 1820 Census of Manufactures*, RG279 (Washington: National Archives, 1944).

59. Historians have debated whether the first successful textile mill was in Massachusetts, Rhode Island, or the Mid-Atlantic. The genealogy of the debate goes back to Robert S. Rantoul, "The First Cotton Mill in America," *Historical Collections of the Essex Institute,* vol. 33 (Salem, MA: Salem Press, 1898); Robert W. Lovett, "The Beverly Cotton Manufactory: Or Some New Light on an Early Cotton Mill," *Bulletin of the Business Historical Society* 26, no. 4 (Dec. 1952): 218–242; and Caroline Ware, *The Early New England Cotton Manufacture, a Study in Industrial Beginnings* (Boston: Houghton Mifflin, 1931), 19; Tucker, *Samuel Slater,* 51.

60. Pitkin, *A Statistical View,* 469.

61. Michael Zakim, "Sartorial Ideologies: From Homespun to Ready-Made," *American Historical Review* 106, no. 5 (Dec. 2001): 1554.

62. Leather was also favored for similar reasons. Peskin, "From Protection to Encouragement," 604.

63. Jeremy, *Henry Wansey and His American Journal*, 61.

64. Clark, *History of Manufactures in the United States,* 208, 373.

65. Jeremy, *Henry Wansey and His American Journal*, 63.

66. Lovett, "The Beverly Cotton Manufactory," 219, 230–235.

67. October 7, 1817, Stockholders' Records, vol. 1, page 42, Boston Manufacturing Co. Collection. Baker Library Historical Collections, Harvard Business School, Boston, MA (hereafter BLHC).

68. These forms of federal support will be discussed in chapter three.

69. Lovett, "The Beverly Cotton Manufactory," 229–231.

70. Entries, January and March 1790. Journal Accounts 1789–1791, Worcester Cotton Manufactory, Business Records of Various New England and New York textile firms, 1710–1938, BLHC.

71. Entry, January 1790. Journal Accounts 1789–1791, Worcester Cotton Manufactory.

72. Entry, November 1791, Journal Accounts 1789–1791, Worcester Cotton Manufactory. For the advertisement, see William R. Bagnall, *The Textile Industries of the United States* (Cambridge, MA: Riverside, 1893), 129. Amounts/ monetary values are cited as they appeared in the original source. Although "dollar" was the standard currency—based on the Spanish piece of eight—variation in currency persisted for decades after the Revolution. For historical exchange rates, see John J. McCusker, *How Much Is That in Real Money? A Historical Commodity Price Index for Use as a Deflator of Money Values in the Economy of the United States*, 2nd ed. (Worcester, MA: American Antiquarian Society, 2001).

73. Bagnall, *Textile Industries*, 130.

74. Arthur H. Cole and Edwin Francis Gay, *Industrial and Commercial Correspondence of Alexander Hamilton: Anticipating His Report on Manufactures* (New York: Augustus M. Kelley, 1968), 78.

75. Jeremy, *Henry Wansey and His American Journal*, 68.

76. Bagnall, *Textile Industries*, 105–108.

77. Cathy D. Matson and Peter S. Onuf, *A Union of Interests: Political and Economic Thought in Revolutionary America* (Lawrence: University Press of Kansas, 1990), 160.

78. Naomi R. Lamoreaux and John Joseph Wallis, "States, Not Nation: The Sources of Political and Economic Development in the Early United States," *Johns Hopkins Institute for Applied Economics, Global Health, and the Study of Business Enterprise, American Capitalism Working Papers*, no. 1 (Sept. 2015).

79. Paul A. Gilje, "The Rise of Capitalism in the Early Republic," *Journal of the Early Republic* 16, no. 2 (Summer 1996): 165. For manufacturers' disappointments with individual state encouragement, see Matson and Onuf, *A Union of Interests*, 160. For the failures of individual states in carrying out internal improvement projects, see John Lauritz Larson, *Internal Improvement:*

National Public Works and the Promise of Popular Government in the Early United States (Chapel Hill: University of North Carolina Press, 2001).

80. Elisha Colt to John Chester, August 20, 1791, in Cole and Gay, *Industrial and Commercial Correspondence*, 11.

81. Peter Colt to John Chester, July 21, 1791, in Cole and Gay, *Industrial and Commercial Correspondence*, 5–6.

82. August 1794 entry, Journal Accounts 1789–1791, Worcester Cotton Manufactory.

83. Thomas Dublin, *Women at Work: The Transformation of Work and Community in Lowell, Massachusetts, 1826–1860* (New York: Columbia University Press, 1981), 15–17.

84. Nor did Slater's mill represent the large-scale industry that would eventually characterize American manufacturing.

85. Slater Family Business Records, Slater, Almy, and Brown, BLHC.

86. Jeremy, *Henry Wansey and His American Journal*, 83.

87. Ben-Atar, *Trade Secrets,* 104; Jeremy, *Transatlantic Industrial Revolution,* 51.

88. Jeremy, *Transatlantic Industrial Revolution*, 101.

89. Jeremy, *Transatlantic Industrial Revolution,* 79; and Tucker, *Samuel Slater,* 50–51.

90. Daybook entry, January 1, 1790, vol. 21, Almy and Brown Records, Mss29, Rhode Island Historical Society, Providence, RI (hereafter RIHS).

91. Tucker, *Samuel Slater,* 57–58.

92. William Brown to Almy and Brown, January 22, 1801; Josiah Townsend to Almy and Brown, March 2, 1801; E. Savage to Almy and Brown, March 10, 1801; and A. and A. Smith to Almy and Brown, April 2, 1801, Unbound Papers 1801–1811, Almy and Brown Records, Mss29, RIHS.

93. Tucker, *Samuel Slater,* 61.

94. A. and A. Smith to Almy and Brown, April 2, 1801; and unsigned letter to Almy and Brown, 1801, Unbound Papers 1801–1811, Almy and Brown Records, Mss29, RIHS.

95. William Brown to Almy and Brown, January 22, 1801, Unbound Papers 1801–1811, Almy and Brown Records, Mss29, RIHS. For women's work in textiles, see Thomas Dublin, "Women's Work and the Family Economy: Textiles and Palm Leaf Hat Making in New England, 1830–1850," *Tocqueville Review* 5 (Fall/Winter 1983): 297–316; and Thomas Dublin, "Women and Outwork in a Nineteenth Century New England Town: Fitzwilliam," in Jonathan Prude and Steven Hahn, eds., *The Countryside in the Age of Capitalist Transformation Essays in the Social History of Rural America* (Chapel Hill: University of North Carolina Press, 1985), 51–70.

96. Unsigned letter from Poughkeepsie to Almy and Brown, 1801, Unbound Papers 1801–1811, Almy and Brown Records, Mss29, RIHS.

97. William Brown to Almy and Brown, March 12, 1801, Unbound Papers 1801–1811, Almy and Brown Records, Mss29, RIHS.

98. Tucker, *Samuel Slater,* 89.

99. Worcester Historical Society, *Landscape of Industry: An Industrial History of the Blackstone Valley* (Lebanon, NH: University Press of New England, 2009), 50.

100. Statistics of Manufacturing of Cotton in Rhode Island in the year 1809, November 8, 1809, Box 1, Folder 1, Correspondence 1805–1819, Zachariah Allen Papers, Ms254, RIHS.

101. Return of North Providence, Pawtucket, US Department of State, *Records of the 1820 Census of Manufactures*, Schedule for Massachusetts and Rhode Island, RG279 (Washington: National Archives, 1944), roll 2.

102. Statistician and statesman Timothy Pitkin lamented, in his compilation of American economic statistics, that "Notwithstanding the short period of the war [of 1812] the American armies suffered severely for the want of blankets and other necessary clothing." Pitkin, *A Statistical View,* 408.

103. Lawrence Peskin, *Manufacturing Revolution: The Intellectual Origins of Early American Industry* (Baltimore: Johns Hopkins University Press, 2007), 119, 98.

104. Gary J. Kornblith, "'Cementing the Mechanic Interest': Origins of the Providence Association of Mechanics and Manufacturers," *Journal of the Early Republic* 8, no. 4 (Winter 1988): 361–362, 372.

105. Jeremy, *Henry Wansey and His American Journal*, 127.

106. Coxe articulated a national vision for an American political economy based on self-sufficiency and territorial growth. Martin Öhman, "Perfecting Independence: Tench Coxe and the Political Economy of Western Development," *Journal of the Early Republic* 31, no. 3 (Fall 2011): 397–433.

107. Mark R. Wilson, *The Business of Civil War: Military Mobilization and the State, 1861–1865* (Baltimore: Johns Hopkins University Press, 2006), 73.

108. Rozann Rothman, "Political Method in the Federal System: Albert Gallatin's Contribution," *Publius* 1, no. 2 (Winter 1972): 132–133.

109. Industry, security, and the economy seemed especially connected as, for example, the United States sold high-quality, American-made warships to Algiers in order to maintain peace and commerce in the Mediterranean. Hannah Farber, "Millions for Credit: Peace with Algiers and the Establishment of America's Commercial Reputation Overseas, 1795–96," *Journal of the Early Republic* 34, no. 2 (Summer 2014): 216.

110. Jeremy, *Henry Wansey and His American Journal*, 65.

111. ASP, Journal of the Senate, 1st Congress, 3rd Session (Dec. 8, 1790): 217.

112. For the Washington administration's planned use of military coercion to suppress domestic rebellion against taxes, see Richard H. Kohn, "The Washington Administration's Decision to Crush the Whiskey Rebellion," *Journal of*

American History 59, no. 3 (Dec. 1972): 567–584. For earlier internal dissent that resulted in military force to quell resistance to economic policies, see David P. Szatmary, *Shays' Rebellion: The Making of an Agrarian Insurrection* (Amherst: University of Massachusetts Press, 1980).

113. Robert E. Shalhope, "The Ideological Origins of the Second Amendment," *Journal of American History* 69, no. 3 (Dec. 1982): 599–614.

114. Lawrence Delbert Cress argues that fears of centralized power led to the celebration of the militia. *Citizens in Arms: The Army and the Militia in American Society to the War of 1812*. (Chapel Hill: University of North Carolina Press, 1982). For the significance of the US defeat in the Northwest Indian Wars, see Colin G. Calloway, *The Victory with No Name: The Native American Defeat of the First American Army* (New York: Oxford University Press, 2014); and William Hogeland, *Autumn of the Black Snake: The Creation of the U.S. Army and the Invasion That Opened the West* (New York: Farrar, Straus and Giroux, 2017).

115. ASP, "Organization of the Militia," Military Affairs 1, no. 2 (Jan. 21, 1790): 12.

116. Robert E. Shalhope, "The Armed Citizen in the Early Republic," *Law and Contemporary Problems* 49, no. 1 (Winter 1986): 131–138. This is not meant to imply that there is or was a "natural" link between gun ownership and American culture; rather, for Americans committed to preserving independence and safeguarding the new nation from perceived threats, firearms were necessary.

117. Coxe, *A brief examination*, 122–123.

118. ASP, "Militia," Military Affairs 1, no. 21 (Dec. 11, 1794): 69; and Don Higginbotham, *The War of American Independence: Military Attitudes, Policies, and Practice, 1763–1789* (Bloomington: Indiana University Press: 1977), 309.

119. ASP, "Militia," Military Affairs 1, no. 23 (Dec. 29, 1794): 107.

120. Annals of Congress, "Arms for the Militia," June 1798, 5th Congress, 2nd Session, 1,927–1,933.

121. Fagal, "The Political Economy of War," 218. Fears of slave rebellion also no doubt prompted southerners to prefer government support of arms production. In 1798, Virginia established a state-owned arms manufactory. Giles Cromwell, *The Virginia Manufactory of Arms* (Richmond: University of Virginia Press, 1975).

122. ASP, "Return of Ordnance, Arms, and Military Stores," Military Affairs 1, no. 10 (Dec. 16, 1793): 44.

123. ASP, "Arsenals and Armories," Military Affairs 1, no. 14 (March 5, 1794): 65.

124. Information in *American State Papers* has military spending at 21 percent of the federal budget in 1794. See "Increase of Duties," *Finance* 1, no. 65 (April 17, 1794): 278. According to "Table Ea636-643—Federal Government

Expenditure, by Major Function: 1789–1970," in Scott Sigmund Gartner, Michael R. Haines, Alan L. Olmstead, Richard Sutch, and Gavin Wright, *Historical Statistics of the United States*, ed. Susan B. Carter (New York: Cambridge University Press, 2006), the army accounted for 37 percent of total expenditures. According to Timothy Pitkin it was 30 percent. See Pitkin, *A Statistical View,* 342.

125. Jeremy, *Henry Wansey and His American Journal*, 65.

126. Michael S. Raber, "Conservative Innovators, Military Small Arms, and Industrial History at Springfield Armory, 1794–1918," *IA: The Journal of the Society for Industrial Archeology* 14, no. 1 (1988): 2.

127. Derwent Stainthorpe Whittlesey, "The Springfield Armory: A Study in Institutional Development" (PhD diss., University of Chicago, 1920), 49.

128. Whittlesey, "The Springfield Armory," 47.

129. "ASP, Military Force, Arsenal, and Stores," Military Affairs 1, no. 25 (Dec. 15, 1795): 108; and James B. Whisker, *The United States Armory at Springfield, 1795–1865* (Lewiston, NY: E. Mellen, 1997), 19.

130. President Washington, in fact, sent a questionnaire to American generals about the military tactics of the French. Whisker, *The United States Armory at Springfield*, 23.

131. ASP, "Armory at Springfield," Military Affairs 1, no. 37 (Jan. 7, 1800): 130.

132. Whittlesey, "The Springfield Armory," 48; and Whisker, *The United States Armory at Springfield,* 18.

133. David Hounshell, *From the American System to Mass Production, 1800–1932: The Development of Manufacturing Technology in the United States,* vol. 4 (Baltimore: Johns Hopkins University Press, 1985), 32.

134. ASP, "Armory at Springfield," Military Affairs 1, no. 37 (Jan. 7, 1800): 130.

135. ASP, "Armory at Springfield," 130.

136. Jeremy, *Henry Wansey and His American Journal*, 70.

137. *The Boston Directory* (Boston: John West, 1800).

138. See, for example, Danvers, Mass., General Store Daybook, 1789–1791, Account Books (unidentified) Collection, 1703–1852, Folio Volume 6, American Antiquarian Society, Worcester, MA (hereafter AAS); and Worcester or Boston, Mass., Wholesale and Imports Account Book, Account Books (unidentified) Collection, 1703–1852, Folio Volume 11, AAS. The controversy over *Arming America's* source fabrication notwithstanding, Michael Bellesiles's argument that an early American gun-owning culture was less pervasive than popular culture suggests is at least partly true. Michael A. Bellesiles, *Arming America: The Origins of a National Gun Culture* (New York: Knopf, 2000).

139. ASP, "Military Force, Arsenal, and Stores," Military Affairs 1, no. 25 (Dec. 15, 1795): 108.

140. ASP, "Causes of the Failure of the Expedition Against the Indians, in 1791, Under the Command of Major General St. Clair," Military Affairs 1, no. 5 (May 8, 1792): 36–38.

141. ASP, "Protection of the Frontier," Military Affairs 1, no. 61 (Jan. 27, 1795): 547; Calloway, *The Victory with No Name*; and Hogeland, *Autumn of the Black Snake*.

142. Anthony Benezet, *Some observations on the situation, disposition, and character of the Indian natives of this continent* (Philadelphia: Joseph Cruckshank, 1784), 58. Political economist and champion of domestic manufactures, Mathew Carey, owned a copy of this. See the copy of Benezet, *Some observations*, at the Library Company of Philadelphia.

143. James V. Joy Jr., "Eli Whitney's Contracts for Muskets," *Public Contract Law Journal* 8, no. 2 (Dec. 1976): 142.

144. ASP, "Suspension of the Recruiting Service," Military Affairs 1, no. 38 (Jan. 13, 1800): 132. Also, in 1796, South Carolina Representative Robert Goodloe Harper acknowledged that many in Congress doubted "the propriety of reducing the number of troops." Robert Goodloe Harper, March 9, 1796, in Noble E. Cunningham Jr., ed., *Circular Letters of Congressmen to Their Constituents, 1789–1829* (Chapel Hill: University of North Carolina Press, 1978), 47.

145. Whisker, *The United States Armory at Springfield*, 23; Samuel Hodgson to John Harris, September 3, 1798, National Archives and Record Administration: Post-Revolutionary War Papers, Record Group 45.

146. David Ames to Samuel Hodgson, March 29, 1799, National Archives and Record Administration: Post-Revolutionary War Papers, Record Group 45.

147. Samuel Hodgson to John Harris, September 3, 1798, National Archives and Record Administration: Post-Revolutionary War Papers, Record Group 45.

148. Drawings and Tables of Foreign Ordnance, vols. 1 and 2, 1787, Records of the Office of the Chief of Ordnance, Record Group 156, Entry 69, National Archives, Washington.

149. Alder, *Engineering the Revolution*, 4.

150. Merritt Roe Smith, *Military Enterprise and Technological Change: Perspectives on the American Experience* (London: MIT Press, 1985), 47–49.

151. For skepticism of Whitney's invention of the cotton gin, see Angela Lakwete, *Inventing the Cotton Gin: Machine and Myth in Antebellum America* (Baltimore: Johns Hopkins University Press, 2005).

152. Jeremy, *Henry Wansey and His American Journal*, 72–73.

153. Oliver Wolcott to Eli Whitney, October 9, 1798, Box 1, Folder 13, Eli Whitney Papers, Yale University Manuscripts and Archives.

154. Smith, *Military Enterprise and Technological Change*, 47–49; Eli Whitney to Oliver Wolcott, October 17, 1798, Box 1, Folder 13, Eli Whitney Papers, Yale University Manuscripts and Archives.

155. Joy, "Eli Whitney's Contracts for Muskets," 144.

156. Oliver Wolcott to Daniel Gilbert, September 8, 1798, National Archives and Record Administration: Post-Revolutionary War Papers, Record Group 45.

157. Wolcott to Gilbert, September 8, 1798.

158. Henry Dearborn to Eli Whitney, February 25, 1803, Order Book, Letters Sent Regarding Procurements, Records of the Office of the Secretary of War, RG107, National Archives, Washington.

159. Decius Wadsworth to Eli Whitney, December 30, 1800, Box 1, Folder 16, Eli Whitney Papers, Yale University Manuscripts and Archives.

160. For the importance of reputation in making economic arrangements work, see Joel Mokyr, "The Institutional Origins of the Industrial Revolution," in Elhanan Helpman, ed., *Institutions and Economic Performance* (Cambridge: Harvard University Press, 2008), 78–79.

161. Eli Whitney to Samuel Dexter, February 7, 1801, Box 1, Folder 17, and Oliver Wolcott to Eli Whitney, October 9, 1798, Box 1, Folder 13, Eli Whitney Papers, Yale University Manuscripts and Archives.

162. Eli Whitney to Oliver Wolcott, June 15, 1798, Box 1, Folder 12, Eli Whitney Papers, Yale University Manuscripts and Archives.

163. Edgar Goodrich to Simeon Baldwin, January 8, 1801, Box 1, Folder 17, Eli Whitney Papers, Yale University Manuscripts and Archives.

164. Eli Whitney to Samuel Dexter January 8, 1801, Box 1, Folder 17, Eli Whitney Papers, Yale University Manuscripts and Archives.

165. Eli Whitney to Oliver Wolcott, July 12, 1798, Box 1, Folder 13, Eli Whitney Papers, Yale University Manuscripts and Archives.

166. "Cost of a Musket," July 21, 1821, Box 8, Folder 114, Eli Whitney Papers, Yale University Archives and Manuscripts; and Robert S. Woodbury, "The Legend of Eli Whitney and Interchangeable Parts," *Technology and Culture* 1, no. 3 (1960): 244–245.

167. The contractors ended up delivering 33,235 muskets of the 36,500 contracted for. Joy, "Eli Whitney's Contracts for Muskets," 45.

168. Oliver Wolcott to Samuel Hodgson, September 20, 1798, National Archives and Record Administration: Post-Revolutionary War Papers, Record Group 45.

169. Whisker, *The United States Armory at Springfield*, 23.

170. Oliver Wolcott to William Henry, December 12, 1799, National Archives and Record Administration: Post-Revolutionary War Papers, Record Group 45.

171. For an example of contract solicitations from the Navy Department to producers of the shipbuilding industry, see "Circular," Boston, June 8, 1812, Amos Binney Letterbook, 1810–1814, AAS.

172. ASP, "Reorganization of the Army," James McHenry to John Adams, Military Affairs 1, no. 35 (Dec. 24 and 31, 1798): 124.

173. ASP, "Reorganization of the Army," 127; James McHenry and Samuel Hodgson to John Harris, September 10, 1798, Coxe Irvine Papers—Philadelphia Supply Agencies: Correspondence, Reports Returns Bill Accounts Receipts Vouchers and Contracts 1794–1842, Box 118, Records of the Office of the Quartermaster General, Record Group 92, National Archives Building, Washington (hereafter RQMG); and James McHenry and Samuel Hodgson to John Harris, August 22, 1799, Coxe Irvine Papers—Philadelphia Supply Agencies: 1800–14 Invoices Orders for clothing and supplies, Box 184, RQMG. Congress had spent over $1.2 million on the navy by the late 1790s as it geared up for war with the Barbary States. Robert J. Allison, *Crescent Obscured: The United States and the Muslim World, 1776–1815* (Chicago: University of Chicago Press, 1995), 30.

174. ASP, "Arsenals and Armories," Military Affairs 1, no. 14 (March 5, 1794): 65.

175. Tench Coxe, "A Statement of the Arts and Manufacture of the United States of America, for the year 1810" (Philadelphia: A. Corman, 1814), ix–xi.

176. Coxe, "A Statement of the Arts and Manufacture," xlvii.

177. Memorandum of goods returned and goods missing on board the Schooner Two Friends, July 11, 1802. Most of the voyages in which Israel Thorndike invested brought in more than thirty or forty thousand dollars. Box 5, Volume 8, 1802, Israel Thorndike Collection, BLHC.

178. Jeremy, *Henry Wansey and His American Journal*, 39.

Chapter 3. Embargo and War

1. J. and S. Hindsdill, Bennington, VT, 1820, US Department of State, *Records of the 1820 Census of Manufactures*, Schedule for Vermont, RG279 (Washington: National Archives, 1944), roll 3.

2. Donald R. Hickey, *The War of 1812: A Forgotten Conflict* (Champaign: University of Illinois Press, 2012); and Alan Taylor, *The Civil War of 1812: American Citizens, British Subjects, Irish Rebels, & Indian Allies* (New York: Vintage, 2010). For the importance of understanding the United States as a war-making power whose engagement in international affairs influenced domestic policies, see Ira Katznelson and Martin Shefter, eds., *Shaped by War and Trade: International Influences on American Political Development* (Princeton: Princeton University Press, 2002). Scholars have pointed to the manufacturing-sector opportunities created by warfare. See, for example, Patrick O'Brien, "The Contributions of Warfare with Revolutionary and Napoleonic France to the Consolidation and Progress of the British Industrial Revolution" (London School of Economics Working Papers 150/11, 2011). For a broader approach to the generative effects of war on economic and industrial growth, see Ronald Findlay and Kevin H. O'Rourke, *Power and Plenty: Trade, War, and the World Economy in the Second Millennium* (Princeton: Princeton University Press, 2007).

3. O'Brien, "The Contributions of Warfare," 2, 48. For the general contraction of the postwar economy, see Douglass Cecil North, *The Economic Growth of the United States, 1790–1860* (Englewood Cliffs, NJ: Prentice-Hall, 1961), 61–62. Although the United States experienced a postwar economic downturn, the scale of warfare and level of debt accumulation were much smaller than what Britain experienced during the Napoleonic Wars and so did not have as significant an impact on economic growth. Jeffrey G. Williamson, "Why Was British Growth So Slow during the Industrial Revolution?" *Journal of Economic History* 44, no. 3 (Sept. 1984): 687–712.

4. Mark R. Wilson, "The Extensive Side of Nineteenth-Century Military Economy: The Tent Industry in the Northern United States during the Civil War," *Enterprise and Society* no. 2 (June 2001): 300–301.

5. Gautham Rao, "The Creation of the American State: Customhouses, Law, and Commerce in the Age of Revolution" (PhD diss., University of Chicago, 2008), 282; and Rao, *National Duties: Customs Houses and the Making of the American State* (Chicago: University of Chicago Press, 2016).

6. ASP, "A Bill, Making Provision for arming and equipping the whole body of the militia of the United States," 10th Congress, 1st Session, no. 90 (April 4, 1808); and Tench Coxe, "A Statement of the Arts and Manufacture of the United States of America, for the year 1810" (Philadelphia: A. Corman, 1814).

7. See, for example, Callendar Irvine, Superintendent of Military Stores, Philadelphia, to Tench Coxe, Purveyor of Public Supplies, Philadelphia, June 5, 1805, Coxe Irvine Papers—Philadelphia Supply Agencies: Correspondence, Reports, Returns, Bill Accounts, Receipts, Vouchers and Contracts 1794–1842, Box 136, RQMG.

8. ASP, "Armory at Springfield," no. 246 (March 3, 1823): 540.

9. ASP, "Firearms," no. 58 (Jan. 11, 1806): 190.

10. George D. Moller, *American Military Shoulder Arms, Volume II: From the 1790s to the End of the Flintlock Period* (Albuquerque: University of New Mexico Press, 2011), chapter 144.

11. Andrew Fagal, "The Political Economy of War in the Early American Republic, 1774–1821" (PhD diss., Binghamton University: State University of New York, 2013), 218.

12. Fagal, "The Political Economy of War," 218; John McLean, Commissary of Military Stores, New York, February 20, 1809, AAS.

13. James B. Whisker, *The United States Armory at Springfield, 1795–1865* (Lewiston, NY: E. Mellen Press, 1997), 31.

14. ASP, "Arming the Militia," April 23, 1808, 10th Congress, 1st Session: 2175.

15. Thomas Mann Randolph, April 27, 1806, in Noble E. Cunningham Jr., ed., *Circular Letters of Congressmen to Their Constituents, 1789–1829* (Published for the Institute of Early American History and Culture by Chapel Hill: University of North Carolina Press, 1978), 478.

16. Whisker, *The United States Armory at Springfield*, 36. All arms manufactured under this act were earmarked for the state militias.

17. ASP, "Estimates of the Expense of Arming and Organizing the Militia of the United States According to a Particular Plan," no. 442 (March 3, 1830): 801.

18. Whisker, *The United States Armory at Springfield*, 40.

19. Whisker, *The United States Armory at Springfield*, 21.

20. "An account of monies advanced by Tench Coxe by order of the War Department for the purpose of procuring arms by contract to manufacture in the United States and for their use," December 31, 1809, Bound Volumes of the Philadelphia Supply Agencies, Volume 13 Manufacture of Clothing, Tailors Work, Contracts 1805–11, RQMG.

21. Starr also manufactured fire engines. James E. Hicks, *Nathan Starr: The First Official Sword and Arms Maker* (New York: James E. Hicks, 1940); and Simeon Newton Dexter North, *Simeon North, First Official Pistol Makers of the United States; A Memoir* (Concord, NH: Rumford Press, 1913), 56.

22. [Tench Coxe?] Rough Draft of Letter to Secretary of War, November 5, 1807, Coxe Irvine Papers—Philadelphia Supply Agencies: Correspondence, Reports, Returns, Bill Accounts, Receipts, Vouchers and Contracts 1794–1842, Box 136, RQMG.

23. "An Account of monies advanced by Tench Coxe by order of the War Department for the purpose of procuring Arms in Contract to Manufacture in the United State for their use," December 3, 1809, Bound Volumes of the Philadelphia Supply Agencies, Volume 13 Manufacture of Clothing, Tailors Work, Contracts 1805–11, RQMG; Whisker, *The United States Armory at Springfield*, 39.

24. Stephen V. Grancsay, "The Craft of the Early American Gunsmith," *The Metropolitan Museum of Art Bulletin* 6, no. 2 (Oct. 1947): 60.

25. "Memorial to John Armstrong," April 3, 1813, and "Sketch Memorial of William and J. J. Henry to Secretary of War," April 20, 1813, Folder 7, Box 8, Henry Family Papers (1209), The Hagley Library and Museum.

26. Donald R. Hoke, *Ingenious Yankees: The Rise of the American System of Manufactures in the Private Sector* (New York: Columbia University Press, 1989), 4.

27. Theodore Steinberg, *Nature Incorporated: Industrialization and the Waters of New England* (Cambridge: Cambridge University Press, 2004).

28. David R. Meyer, "Formation of Advanced Technology Districts: New England Textile Machinery and Firearms, 1790–1820," *Economic Geography* 74, no. 1 (Mar. 1998): 31; and David R. Meyer, *Networked Machinists: High-Technology Industries in Antebellum America* (Baltimore: Johns Hopkins University Press, 2006).

29. Whisker, *The United States Armory at Springfield*, 18; and Meyer, "Formation of Advanced Technology Districts," 40.

30. Meyer, "Formation of Advanced Technology Districts," 40.

31. "An Act for the Relief of Adam Kinsley, Thomas French, and Charles Leonard," February 20, 1819, in Richard Peters, *By Authority of Congress, The Public Statutes at Large of the United States of American from the Organization of the Government in 1789 to March 3, 1845* (Boston: Little, Brown, 1846), 223.

32. Fagal, "The Political Economy of War," 235.

33. Coxe, "A Statement of the Arts and Manufacture."

34. Gunsmith Manufactory for Small Arms, Pittsfield, Massachusetts, December 30, 1820, US Department of State, *Records of the 1820 Census of Manufactures*, Schedules for Massachusetts and Rhode Island, RG279 (Washington: National Archives, 1944), roll 2.

35. North, *Simeon North, First Official Pistol Makers*, 76.

36. ASP, "An Act to prohibit the importation of certain goods, wares and merchandise," April 18, 1806, 379.

37. J. C. A. Stagg, *Mr. Madison's War: Politics, Diplomacy, and Warfare in the Early American Republic, 1783–1830* (Princeton: Princeton University Press, 1983), 22.

38. Caroline F. Ware, "The Effect of the American Embargo, 1807–1809, on the New England Cotton Industry," *Quarterly Journal of Economics* 40, no. 4 (Aug. 1926): 684.

39. Edward E. Baptist, *The Half Has Never Been Told: Slavery and the Making of American Capitalism* (New York: Basic, 2014), 322.

40. Anthony E. Kaye, "The Second Slavery: Modernity in the Nineteenth-Century South and the Atlantic World," *Journal of Southern History* 75, no. 3 (Aug. 2009): 627–650.

41. Baptist, *The Half Has Never Been Told*; Adam Rothman, *Slave Country: American Expansion and the Origins of the Deep South* (Cambridge: Harvard University Press, 2005). And early in the nineteenth century southern planters still saw the Union in terms of complementary economic interests by which the plantation south could produce cotton for an increasingly industrial north. Brian Schoen, *The Fragile Fabric of Union: Cotton, Federal Politics, and the Global Origins of the Civil War* (Baltimore: Johns Hopkins University Press, 2009).

42. Ware, "The Effect of the American Embargo," 684–685. For the firm's expansion into new markets, see Burr Woolman and Co. [Burlington] to Almy and Brown, September 20, 1808, Unbound Papers 1801–1811, Almy and Brown Records, Mss29, RIHS.

43. Ware, "The Effect of the American Embargo," 682.

44. Burr Woolman and Co. [Burlington] to Almy and Brown, September 20, 1808, Unbound Papers 1801–1811, Almy and Brown Records, Mss29, RIHS.

45. US Department of State, *Records of the 1820 Census of Manufactures*, Schedule for Vermont, RG279 (Washington: National Archives, 1944).

46. ASP, "Organization of the Militia," Military Affairs 1, no. 2 (Jan. 21, 1790): 11.

47. For a list of some of the tailors employed by the government, see "Memorandum of the quantity of cloth taken by each taylor separately in making the Army Clothing for the year 1804," Coxe Irvine Papers—Philadelphia Supply Agencies: Correspondence, Reports, Returns, Bill Accounts, Receipts Vouchers, and Contracts 1794–1842, Volume 13 Box 145, RQMG.

48. See, for example, "Invoice of Madrid, deposited in hands of Silas E. Weir," Philadelphia, October 12, 1810, Auction Accounts of Weir, Willing, Lewis, Smith and Lisle, Auctioneers Oct 1810–Sept 1811, Folder 1 Silas E. Weir Auction Accounts, October 1810, Mrs. Howard W. Lewis Collection of Early Philadelphia Business Papers, 1799–1866 #0367, HSP.

49. Scott Reynolds Nelson, *A Nation of Deadbeats: An Uncommon History of America's Financial Disasters* (New York: Knopf, 2012), 54.

50. See, for example, the purchase of cloth for military uniforms from Bulmer, Horner and Co. of Leeds, Great Britain, Memorandum: Purveyor of Public Supplies' Office, Philadelphia, June 28, 1806, Volume 13 Box 145, RQMG.

51. Stagg, *Mr. Madison's War,* 173.

52. Reneé Chartrand, *"A Most Warlike Appearance": Uniforms, Flags, and Equipment of the United States Forces in the War of 1812* (Ottawa, ON, Canada: Service Publications, 2011), 27–28.

53. Stagg, *Mr. Madison's War,* 157.

54. Fagal, "The Political Economy of War," 212.

55. See, for example, *Salem Gazette*, November 25, 1808, which lamented the dire commercial conditions and advocated the encouragement of domestic manufactures; "American Manufactured Army clothing," July 6, 1811, Purveyors Office in [Boston] *Independent Chronicle*, July 15, 1811, which advertised that "proposals will be accepted at this office." Also, ads in [New York] *Public Advertiser,* July 22, 1811, solicited green and blue broad cloth, twilled woolen blankets, and white woolens.

56. Annals of Congress, March 3, 1812, 12th Congress, 1st Session, vol. 23: 161.

57. Almy and Brown to Tench Coxe, March 15, 1809, vol. 111, Letterbook 1808–1810, Almy and Brown Records, Mss29, RIHS.

58. Tench Coxe, Purveyor of Public Supplies, Philadelphia, to The Cotton Manufacturers of the United States, July 27, 1810, Unbound Papers 1801–1811, Almy and Brown Records, Mss29, RIHS.

59. Donald H. Kagin, "Monetary Aspects of the Treasury Notes of the War of 1812," *Journal of Economic History* 44, no. 1 (Mar. 1984): 71.

60. Stagg, *Mr. Madison's War,* 158; Smith, *Military Enterprise and Technological Change,* 43.

61. Stagg, *Mr. Madison's War,* 160, 171.

62. Kagin, "Monetary Aspects of the Treasury Notes," 75.

63. Kagin, "Monetary Aspects of the Treasury Notes," 85.

64. The increase in troops was not easily achieved, as many American men preferred independent labor to the low pay and five-year service term of army life. Taylor, *The Civil War of 1812*, 325.

65. Whisker, *The United States Armory at Springfield*, 41.

66. Karim M. Tiro, "The View from Piqua Agency: The War of 1812, the White River Delawares, and the Origins of Indian Removal," *Journal of the Early Republic* 35, no. 1 (Spring 2015): 28.

67. ASP, "Armories and Arms Manufactured Therein," Military Affairs 1, no. 109 (Feb. 17, 1812): 317.

68. Merritt Roe Smith, *Harpers Ferry Armory and the New Technology: The Challenge of Change* (Ithaca: Cornell University Press, 1977), 170.

69. ASP, "National Armories," Military Affairs 1, no. 141 (Feb. 20, 1815): 604.

70. Whisker, *The United States Armory at Springfield*, 41.

71. See, for example, Callendar Irvine, Commissary General's Office, to John Armstrong, Secretary of War, March 1, 1814, in James E. Hicks, *United States Ordnance: Volume II, Ordnance Correspondence Relative to Muskets, Rifles, Pistols and Swords* (New York: James E. Hicks, 1940), 40. Irvine complained of R. and C. Leonard's failure to comply with the terms of their contract.

72. *Ordnance Reports* 1: 176–177, in Whisker, *The United States Armory at Springfield*, 39.

73. "An Account of monies advanced by Tench Coxe by order of the War Department for the purpose of procuring Arms in Contract to Manufacture in the United State for their use," December 3, 1809, Bound Volumes of the Philadelphia Supply Agencies, Volume 13 Manufacture of Clothing, Tailors Work, Contracts 1805–1811, RQMG; Moller, *American Military Shoulder Arms*, vol. 2, 230.

74. Felicia Johnson Deyrup, *Arms Makers of the Connecticut Valley: A Regional Study of the Economic Development of the Small Arms Industry, 1798–1870* (Northampton, MA, 1948), 47–48; and *Ordnance Reports* 1: 176–177, in Whisker, *The United States Armory at Springfield*, 48.

75. Deyrup, *Arms Makers of the Connecticut Valley*, 45.

76. See Waters Family, Papers, Octavo vol. 1 and Boxes W Box 1, Folder 3, AAS.

77. "An Account of monies advanced by Tench Coxe by order of the War Department for the purpose of procuring Arms in Contract to Manufacture in the United State for their use," December 3, 1809, Bound Volumes of the Philadelphia Supply Agencies, Volume 13 Manufacture of Clothing, Tailors Work, Contracts 1805–1811, RQMG.

78. Asa Waters to Roswell Lee, September 3, 1817, Box 3, Folder 1, Letters Received Miscellaneous, Records of the Springfield Armory, MA, Record Group 156, Entry 1362, NM-59, 94-066, National Archives Building, Waltham, MA (hereafter SA-LRM).

79. Joel Mokyr, "The Institutional Origins of the Industrial Revolution," in Elhanan Helpman, ed., *Institutions and Economic Performance* (Cambridge: Harvard University Press, 2008).

80. David Hounshell, *From the American System to Mass Production, 1800–1932: The Development of Manufacturing Technology in the United States*, vol. 4 (Baltimore: Johns Hopkins University Press, 1985), 32.

81. Hounshell, *From the American System to Mass Production,* 32.

82. Henry Dearborn to Eli Whitney, February 25, 1803, Order Book, Letters Sent Regarding Procurements, Records of the Office of the Secretary of War, Record Group 107, Entry 10, National Archives, Washington.

83. See, for example, Oliver Wolcott, "Bills Receivable in July 1810," Oliver Wolcott Account books, 1781–1831, Connecticut Historical Society, Hartford (hereafter CHS); Hounshell, *From the American System to Mass Production,* 32.

84. Joy, "Eli Whitney's Contracts for Muskets," 150–151.

85. Joy, "Eli Whitney's Contracts for Muskets," 141.

86. North, *Simeon North, First Official Pistol Makers*, 79.

87. Simeon North, December 1826, Firearms Makers Collection, Box 1, Folder 1, Middlesex County Historical Society (hereafter MCHS).

88. Simeon North, Middletown, CT, US Department of State, *Records of the 1820 Census of Manufactures*, Schedule for Connecticut, RG279 (Washington: National Archives, 1944), roll 4.

89. North, *Simeon North, First Official Pistol Makers*, 78–79, 86.

90. Deyrup, *Arms Makers of the Connecticut Valley,* 88.

91. Ken Alder, *Engineering the Revolution: Arms and Enlightenment in France, 1763–1815* (University of Chicago Press, 2010), 131, 223, 249.

92. Brian Balogh writes that following the War of 1812, US leaders moved to create a "more assertive General Government" that would increase funding for defense and lead to military preparedness. Balogh, *A Government out of Sight: The Mystery of National Authority in Nineteenth Century America* (Cambridge: Cambridge University Press, 2009), 202.

93. John Morton to Roswell Lee, July 24, 1816, Box 1, Folder 7, SA-LRM.

94. Benjamin Moore to William Eustis, Secretary of War, November 24, 1812; Moore to Eustis, June 30, 1813; Moore to John Chaffee, Paymaster and Storekeeper, Springfield Armory, May 1, 1813; and Benjamin Moore, Letters about Guns, 1812–1813, MS 101435, CHS.

95. John Morton to Roswell Lee, November 14, 1817, Box 1, Target #2, Letters Received from Officials and Officers of the War and Treasury Depart-

ments, Records of the Springfield Armory, MA, Record Group 156, Entry 1362, NM-59, 94-066; National Archives Building, Waltham, MA (hereafter SA-LRO); and John Morton to Roswell Lee, July 24, 1816, Box 1, Folder 7, SA-LRM.

96. Deyrup, *Arms Makers of the Connecticut Valley,* 57.

97. John Morton to Roswell Lee, March 4, 1818, Box 1, Target #2; George Bomford to Roswell Lee, June 1, 1823, Box 2, SA-LRO.

98. Decius Wadsworth to Roswell Lee February 22, 1816, Box 2, Folder 5, SA-LRM.

99. John Morton to Roswell Lee, July 16, 1816, Box 1, Folder 7, SA-LRM.

100. John Morton, Ordnance Department, to Roswell Lee, July 24, 1816, Box 1, Folder 7, SA-LRM.

101. Deyrup, *Arms Makers of the Connecticut Valley,* 58.

102. ASP, "An Act to prohibit American vessels from proceeding to or trading with the enemies of the United States and for other purposes," no. 129 (July 6, 1812): 778.

103. Ware, "The Effect of the American Embargo."

104. Arnold Manufacturing Company Account Books, CHS.

105. It is difficult to procure exact numbers because many textile records do not survive, and the 1820 manufacturing census is far from complete. Additionally, manufacturers provided varying amounts of information and many did not refer to the founding date of their factories. Arnold Manufacturing Company in Woodstock, Connecticut, for example, was established in 1814, but omitted that information from their response to the census. US Department of State, *Records of the 1820 Census of Manufactures*, Schedule for Connecticut, RG279 (Washington: National Archives, 1944), roll 4; Arnold Manufacturing Company Account Books, CHS. Additionally, during the war, observers noted between twenty-five and forty-five new factories within thirty miles of Providence. Caroline Ware, *The Early New England Cotton Manufacture, a Study in Industrial Beginnings* (Boston: Houghton Mifflin, 1931) 56.

106. Amos Binney to Secretary of Navy William Jones, June 9, 1813, Amos Binney Letterbook, 1810–1814, AAS.

107. ASP, "Clothing the Army with Domestic Fabrics," Military Affairs 2, no. 180 (Jan. 11, 1820): 42.

108. See, for example, *Relf's Philadelphia Gazette*, November 5 to December 17, 1814, p.1.

109. Ware, *The Early New England Cotton Manufacture,* 55.

110. Slater and Tiffany to Jeremiah Brown, March 11, 1816, Vol. H63 Consignments, Slater and Tiffany, Slater Family Business Records, BLHC.

111. Almy and Brown to Archibald Sturbuch, June 23, 1813, and Almy and Brown to Daniel Waldo, June 19, 1813, Vol. 112, Letterbook 1813–1815, Almy and Brown Records, Mss29, RIHS.

112. Almy and Brown to Daniel Waldo, June 21, 1813, Vol. 112, Letterbook 1813–1815, Almy and Brown Records, Mss29, RIHS.

113. Philip Scranton, *Proprietary Capitalism: The Textile Manufacture at Philadelphia, 1800–1885* (London: Cambridge University Press, 1983), 76.

114. Philadelphia merchant David Lapsley, for example, sold drab cloth to the federal government for the Lewis and Clark expedition in 1803. Lapsley Family Business Records (McA MSS 008), McAllister Collection, the Library Company of Philadelphia.

115. Scranton, *Proprietary Capitalism,* 108.

116. Koistinen, *Beating Plowshares into Swords*, 64; Chartrand, "*A Most Warlike Appearance*," 38.

117. Huntington & Backus Woolen Manufactory, Norwich, CT. US Department of State, *Records of the 1820 Census of Manufactures*, Schedule for Connecticut, RG279 (Washington: National Archives, 1944), roll 4.

118. Victor S. Clark, *History of Manufactures in the United States* (Washington: Carnegie Institute, 1929), 568.

119. Scranton, *Proprietary Capitalism,*108.

120. Barbara M. Tucker, *Samuel Slater and the Origins of the American Textile Industry, 1790–1860* (Ithaca: Cornell University Press, 1984), 33.

121. Receipt, December 27, 1810, Box 6, Francis Cabot Lowell Papers, Massachusetts Historical Society, Boston, MA (hereafter MHS).

122. Francis Cabot Lowell to William Cabot, May 28, 1811, Lowell to Ann Grant, May 1811, Box 6, Francis Cabot Lowell Papers, MHS.

123. Nathan Appleton, *Introduction of the Power Loom, and Origin of Lowell* (Lowell, MA: B. H. Penhallow, 1858), 7.

124. January 28, 1813, Stockholders' Records, Volume 1, page 11, Boston Manufacturing Co. Collection, BLHC.

125. ASP, "Clothing the Army with Domestic Fabrics," Military Affairs 2, no. 180 (Jan. 11, 1820): 42.

126. For these men's business ventures and extensive personal and political connections, see Robert F. Dalzell, *Enterprising Elite: The Boston Associates and the World They Made* (New York: W. W. Norton, 1987). Some of these connections will be discussed in greater detail in chapter 4. For the ways in which these connections functioned in banking and made possible both their own financial undertakings and economic development more generally, see Naomi R. Lamoreaux, *Insider Lending: Banks, Personal Connections, and Economic Development in Industrial New England* (New York: Cambridge University Press, 1996). For diplomatic connections, see, for example, George Erving to Israel Thorndike, July 15, 1802, Box 5, Volume 8, Israel Thorndike Collection, BLHC; and Thomas Appleton and Co. to S.K. Jones and Co., August 1, 1815, Box 3, Folder 2, Appleton Family Papers, MHS.

127. James Russell to Francis Cabot Lowell, September 1, 1812, Francis Cabot Lowell Papers, MHS.

128. Lindsay Schakenbach, "From Discontented Bostonians to Patriotic Industrialists: The Boston Associates and the Transcontinental Treaty, 1790–1825," *New England Quarterly* 84 (Sept. 2011): 383–388. Shipping claims will be discussed in more detail in chapter 4.

129. See, for example, the response of John Arms, Conway, Massachusetts, US Department of State, *Records of the 1820 Census of Manufactures*, Schedule for Massachusetts and Rhode Island, RG279 (Washington: National Archives, 1944), roll 2; and October 29, 1813, December 20, 1813, Directors' Records, Volume 2, Boston Manufacturing Co. Collection, BLHC. Lamoreaux, *Insider Lending.*

130. February 7 and July 17, 1815, Directors' Records, Volume 2, Boston Manufacturing Co. Collection, BLHC.

131. November 6, 1815, Directors' Records, Volume 2, Boston Manufacturing Co. Collection, BLHC.

132. Annals of Congress, "Treaty of Peace," February 18, 1815, 13th Congress, 3rd Session, 255–256.

133. Amos Binney to Tobias Lord, January 6, 1814, and Amos Binney to Isaac Hull, January 6, 1814, Amos Binney Letterbook, AAS.

134. Frederick Wolcott, Litchfield County, Connecticut, US Department of State, *Records of the 1820 Census of Manufactures*, Schedule for Connecticut, RG279 (Washington: National Archives, 1944), roll 4.

135. Cotton Factory in Coventry, Rhode Island, US Department of State, *Records of the 1820 Census of Manufactures*, Schedule for Massachusetts and Rhode Island, RG279 (Washington: National Archives, 1944), roll 2.

136. Koistinen, *Beating Plowshares into Swords,* 64; Chartrand, *"A Most Warlike Appearance,"* 38.

137. For the postwar trade in British "monkey" jackets, see Nelson, *A Nation of Deadbeats*, 52–54.

138. Samuel Ogden, *Thoughts, what probable effect the peace with Great Britain will have on cotton manufactures of this country* (Providence, RI: Goddard & Mann, 1815), 13.

139. William Richardson to Francis Cabot Lowell, July 7, 1815, Box 8, Folder 5, Francis Cabot Lowell Papers, MHS; Nathan and Charles Appleton to Thomas Jones and Co., March 15, 1815, Box 3, Folder 1, Appleton Family Papers, MHS; and Jona Morrill and Son to Jeremiah Nelson, Volume 4, Records Collection of the Office of Naval Records and Library, Record Group 25, Entry 328, National Archives Building, Washington.

140. Coxe, "A Statement of the Arts and Manufacture." "An Account of monies advanced by Tench Coxe by order of the War Department for the purpose of procuring Arms in Contract to Manufacture in the United State for their

use," December 3, 1809, Bound Volumes of the Philadelphia Supply Agencies, Volume 13 Manufacture of Clothing, Tailors Work, Contracts 1805–1811, RQMG; and Stephen Jenks and Sons Cotton Manufactory, US Department of State, *Records of the 1820 Census of Manufactures*, Schedule for Massachusetts and Rhode Island, RG279 (Washington: National Archives, 1944), roll 2.

141. Annals of Congress, "Arming the Militia of the West," Letter of Colonel Bomford, January 8, 1823, 17th Congress, 2nd Session, 1314.

142. ASP, "Western Armory," Military Affairs 1, no. 166 (Dec. 7, 1818): 773.

143. ASP, "Western Armory," 773.

144. Smith, *Harpers Ferry Armory and the New Technology*, 59–61.

145. Response of Lemuel Pomeroy, Pittsfield, Massachusetts, US Department of State, *Records of the 1820 Census of Manufactures*, Schedule for Massachusetts and Rhode Island, RG279 (Washington: National Archives, 1944), roll 2.

146. Nathan Starr Sr. to Nathan Starr Jr., September 5, 1815, Box 1, Folder 1, Arms Correspondence, Starr Family Collection, MCHS.

147. John and Mary Lou Missall, *The Seminole Wars: America's Longest Indian Conflict* (Gainesville, FL: University Press of Florida, 2004), 22. For the coercion and negotiation involved in the removal of Creek Indians from the Southeast before the more well-known Cherokee Trail of Tears, see Christopher D. Haveman, *Rivers of Sand: Creek Indians Emigrations, Relocation, and Ethnic Cleansing in the American South* (Lincoln: University of Nebraska Press, 2016).

148. ASP, "Re-Organization of the Militia," Military Affairs 1, no. 152 (Jan. 7, 1817): 663.

149. Annals of Congress, Senate Proceedings, December 2, 1818, 33.

150. Amasa Stetson, Deputy Commissary's Office, Boston, to Callendar Irvine, December 8, 1812, Box 136, RQMG.

151. ASP, "Clothing the Army with Domestic Fabrics," Military Affairs 2, no. 180 (Jan. 11, 1820): 42. This "perfection" also, no doubt, reflected the demands of southern plantation owners for slave goods produced in northern factories. Seth Rockman, *Plantation Goods and the National Economy of Slavery in Antebellum America* (Chicago: University of Chicago Press, forthcoming).

152. ASP, "Report of the Committee on the Subject of Clothing the Militia," no. 38 (Jan. 12, 1820).

Chapter 4. Financing Industry through Florida

1. Gregory Evans Dowd, *A Spirited Resistance: The North American Indian Struggle for Unity, 1745–1815* (Baltimore: Johns Hopkins University Press, 1992).

2. Edward Baptist, *Creating an Old South: Middle Florida's Plantation Frontier before the Civil War* (Chapel Hill: University of North Carolina Press, 2002), 39.

3. Max M. Edling has argued that the United States created a revenue system that allowed it to effectively mobilize resources for war and achieve dominance on the North American continent. Max Edling, "War, Money, and the American States, 1783–1867," paper presented at "Political Economy and Empire in the Early Modern World" conference at Yale University, May 4, 2013.

4. For a recent analysis of the importance of Florida for Manifest Destiny, see Laurel Clark Shire, *The Threshold of Manifest Destiny: Gender and National Expansion in Florida* (Philadelphia: University of Pennsylvania Press, 2016). Shire focuses on the role women played physically and symbolically in settling Florida.

5. February 28, 1820 entry, Volume 2 Director's Records, p. 227, Boston Manufacturing Company Collection. BLHC.

6. Chaim M. Rosenberg, *Life and Times of Francis Cabot Lowell, 1775–1817* (Lanham, MD: Lexington Books, 2011), 252–253.

7. For example, the Monroe Administration was always willing to forgo acquisition of Texas, which other scholars have pegged as the site of American imperial and racial ambitions during the 1830s and 1840s, in its negotiations with Spain for Florida. William Weeks, *John Quincy Adams and American Global Empire* (Lexington: University Press of Kentucky, 1992), 167.

8. Drew McCoy, *The Elusive Republic: Political Economy in Jeffersonian America* (Chapel Hill: University of North Carolina Press, 1980), 198–199. Martin Öhman argues that early republican policy makers understood that territorial expansion was linked with economic security and an improved status in the international trading system. The Louisiana Purchase, in fact, was executed with the understanding that the manufacturing sector would subsequently expand. Öhman, "Perfecting Independence: Tench Coxe and the Political Economy of Western Development," *Journal of the Early Republic* 31, no. 3 (Fall 2011): 397–433.

9. Edward E. Baptist, *The Half Has Never Been Told: Slavery and the Making of American Capitalism* (New York: Basic, 2014), 112.

10. For border fortifications and personnel, see ASP, "Military Force," Military Affairs 1, no. 55 (Feb. 15, 1805): 175. For spending on Indian relations, see Timothy Pitkin, *A Statistical View of the Commerce of the United States of America: Its Connection with Agriculture and Manufactures: and an Account of the Public Debt, Revenues, and Expenditures of the United States* (James Eastburn, 1817), 342, and Pitkin, *A Statistical View of the Commerce of the United States of America: Including also an Account of Banks, Manufactures and Internal Trade and Improvements* (New Haven: Durrie & Peck: 1835), 342.

11. Weeks, *John Quincy Adams and American Global Empire*, 26–27, 30–33.

12. For a brief account of this rebellion, see Walter Johnson, *River of Dark*

Dreams: Slavery and Empire in the Cotton Kingdom (Cambridge: Harvard University Press, 2013), 18–22.

13. This pursuit had been interrupted somewhat during the War of 1812. John and Mary Lou Missall, *The Seminole Wars: America's Longest Indian Conflict* (Gainesville: University Press of Florida, 2004), 20. For an account of the fighting in Florida during the War of 1812, see, for example, Arsène Lacarrière Latour, *Historical memoir of the war in West Florida and Louisiana in 1814–15* (Philadelphia: John Conrad and J. Maxwell, 1816).

14. Although little studied, the First Seminole War was a transformative event in early national history. Deborah Rosen has recently argued that it was foundational in establishing US policies on territoriality, war making, and foreign relations. Deborah A. Rosen, *Border Law: The First Seminole War and American Nationhood* (Cambridge: Harvard University Press, 2015).

15. Thomas Cochran, "The Business Revolution," *American Historical Review* 79, no. 5 (Dec. 1974): 1449–1466.

16. Alfred D. Chandler, *The Visible Hand: The Managerial Revolution in American Business* (Cambridge: The Belknap Press of Harvard University Press, 1977), 71–74.

17. These duties ranged from 15 to almost 40 percent. "Duties payable by law on all goods, wares and merchandise, imported into the United States of America, after the last day of June 1794" (Philadelphia, 1794); US Treasury Department, "A Correct List of Duties" (New York: Longworth, 1806); and "Duties payable by law on all goods, wares and merchandise imported into the United States of America, after the last day of June, eighteen hundred and twelve" (Washington, 1812). Mark Wilson has shown how military supply bureaus became the most stable and bureaucratic arm of the US government in the antebellum years. Mark R. Wilson, *The Business of Civil War: Military Mobilization and the State, 1861–1865* (Baltimore: Johns Hopkins University Press, 2006), 35.

18. Joshua Aubin, "A Few Particulars in Relation to Woolen Manufacturing During an Agency in Amesbury of 31 years commencing in 1821 and ending in 1852," 1854, BLHC.

19. "An Act to Incorporate the Merrimack Manufacturing Company," Vol. 2 Stockholders' Minutes, Merrimack Manufacturing Co. Collection, BLHC.

20. Economic historian Vera Shlakman gave them this name in the 1930s. Vera Shlakman, *Economic History of a Factory Town: A Study of Chicopee, Massachusetts* (Northampton, MA: Department of History, Smith College, 1935).

21. Rosenberg, *Life and Times of Francis Cabot Lowell*, 257.

22. Douglass North, *The Economic Growth of the United States, 1790–1860* (Englewood Cliffs, NJ: Prentice-Hall, 1961), 165, 178, 185.

23. Response of Cotton Manufactory in Winthrop, Kennebunk, Maine, Nov. 20, 1820, US Department of State, *Records of the 1820 Census of*

Manufactures, Schedule for Maine and New Hampshire, Microcopy number 279, National Archives, Washington, 1964, roll 1; and Edward Stanwood, *American Tariff Controversies in the Nineteenth Century*, vol. 1 (Boston: Houghton, Mifflin, 1903), 160. The duties on cotton thread, yarn, and cloths, for example, increased from 25 to 35 percent.

24. Embajada, Spain, (US) *Official correspondence between Don Luis de Onis and John Quincy Adams: in relation to the Floridas and the boundaries of Louisiana, with other matters in dispute between the two governments* (London: E. Wilson, 1818).

25. Petitions from Chesapeake Insurance Co., the Union Insurance Co., the Maryland Insurance Co., and the Baltimore Insurance Co. sent to Congress in 1826, *Senate Committee on Foreign Relations,* Jan. 3, 1826, and Mar. 1, 1826 (SEN19A-G6), National Archives, Washington.

26. US Department of State, Library of Congress, *Treaty of Amity, Settlement, and Limits between the United States of America and His Catholic Majesty, 1819* (Washington: GPO, 1909), available at http://avalon.law.yale.edu/19th_century/sp1819.asp.

27. Alfred S. Konefsky and Andrew J. King, eds., *The Papers of Daniel Webster, Legal Papers,* vol. 2: *The Boston Practice* (Hanover, NH: University Press of New England, 1983), 228.

28. See, for example, Robert F. Dalzell, *Enterprising Elite: The Boston Associates and the World They Made* (New York: W. W. Norton, 1987); Alfred D. Chandler, *The Visible Hand: The Managerial Revolution in American Business* (Cambridge: Belknap Press of Harvard University Press, 1977); Frances W. Gregory, *Nathan Appleton: Merchant and Entrepreneur, 1779–1861* (Charlottesville: University of Virginia Press, 1975); Chaim M. Rosenberg, *Life and Times of Francis Cabot Lowell, 1775–1817* (Lanham, MD: Lexington Books, 2011); and Carl Seaburg and Stanley Paterson, *Merchant Prince of Boston: Colonel T. H. Perkins, 1764–1854* (Cambridge: Harvard University Press, 1971).

29. Edward Stanwood, *American Tariff Controversies in the Nineteenth Century*, vol. 1 (Boston: Houghton, Mifflin, 1903), 202–203.

30. John Lauritz Larson, "'Bind the Republic Together': The National Union and the Struggle for a System of Internal Improvements," *Journal of American History* 74, no. 2 (Sept. 1987): 386–387; and Samuel Flagg Bemis, *John Quincy Adams and the Union* (New York: Knopf, 1956), 89.

31. Weeks, *John Quincy Adams and American Global Empire*, 164.

32. The merchants from Philadelphia and Baltimore, for example, petitioned the Senate Committee on Foreign Relations for insufficient claims payments; the Associates did not. See, for example, petitions by Philadelphia and Baltimore sent to Congress in 1826, *Senate Committee on Foreign Relations,* Jan. 3, 1826. and Mar. 1, 1826 (SEN19A-G6), National Archives, Washington.

33. Daniel Webster to Joseph Hopkinson, February 1, 1822, in Konefsky and King, eds., *The Papers of Daniel Webster, Legal Papers,* vol. 1, 305.

34. Charles Eisen's recent work on John Quincy Adams reveals the leader as a shrewd politician and pragmatic statesman in his crafting of American policies that would expand the nation's borders, support manufacturing, and mitigate foreign threats. Charles N. Eisen, *Nation Builder: John Quincy Adams and the Grand Strategy of the Republic* (Cambridge: Harvard University Press, 2014).

35. See, for example, "Six percent stock of 1813," October 14, 1816, Box 3, Folder 4, General Correspondence, *Appleton Family Papers*, MHS.

36. Gregory, *Nathan Appleton*, 100. Merchants also used St. Mary's, Georgia, for illicit trade. Christopher Ward, "The Commerce of East Florida during the Embargo, 1806–1812: The Role of Amelia Island," *Florida Historical Quarterly* 68, no. 2 (Oct. 1989): 179. T. H. Perkins, who would receive a significant share of the Spanish Claims payouts, traded slaves until around 1810, two years after the slave trade's official end. N. P. Russell to T. H. and James Perkins, Jan. 25, 1810, *Thomas Handasyd Perkins Papers*, MHS. The Perkins Company, which T. H. inherited after his father's death, had been heavily involved in the slave trade in the West Indies, particularly in Cape Francis, San Domingo. Jane Thompson-Stahr, *The Burling Books: Ancestors and Descendants of Edward and Grace Burling, Quakers, 1600–2000* (Louisville, KY: Gateway, 2001), 315–316.

37. Weeks, *John Quincy Adams and American Global Empire,* 52.

38. Weeks, *John Quincy Adams and American Global Empire*, 48–52.

39. Harrison Gray Otis to Sally Foster Otis, February 28, 1820, Harrison Gray Otis Papers, MHS.

40. Konefsky and King, eds., *The Papers of Daniel Webster*, 183.

41. Peter Chardon Brooks to William King, June 4, 1821, and Aug. 3, 1821, in Konefsky and King, eds., *The Papers of Daniel Webster.*

42. Daniel Webster to Alexander Bliss, March 2, 1822, in Charles M. Wiltse, ed., *Microfilm Edition of the Papers of Daniel Webster* (Hanover, NH: Dartmouth College, 1975), Reel 3.

43. Dalzell, *Enterprising Elite,* 28–29.

44. Robert Vincent Remini, *Daniel Webster: The Man and His Time* (New York: W. W. Norton, 1997), 138, 146; and Gregory, *Nathan Appleton,* 127.

45. Accounts Current, July 1822, Volume 35, Boston Manufacturing Company Records, BLHC.

46. Daniel Webster to Samuel Jaudon, June 5, 1822, Charles M. Wiltse, ed. *Microfilm Edition of the Papers of Daniel Webster,* Reel 3.

47. Daniel Webster to Alexander Bliss, Jan. 4, 1824, in Konefsky and King, eds., *The Papers of Daniel Webster.*

48. Application to Postpone Cases, July 21, 1823, and D. W. to Alexander

Bliss, Dec. 30, 1822, in Konefsky and King, eds., *The Papers of Daniel Webster,* 219–221.

49. For examples of the nature of the Associates' trade in the West Indies, see Gregory, *Nathan Appleton*, 35, 38; J. D. Forbes, *Israel Thorndike, Federalist Financier* (New York: Exposition Press, 1953), 29–30; and Thompson-Stahr, *The Burling Books*, 315–316.

50. N. P. Russell to T. H. and James Perkins, Jan. 25, 1810, *Thomas Handasyd Perkins Papers*, MHS. Thompson-Stahr, *The Burling Books*, 477.

51. Seaburg and Paterson, *Merchant Prince of Boston*, 395.

52. Dalzell, *Enterprising Elite*, 133–134.

53. Carl E. Prince and Seth Taylor, "Daniel Webster, the Boston Associates, and the U.S. Role in the Industrializing Process, 1815–1830," *Journal of the Early Republic* 2, no. 3 (Autumn 1982): 294.

54. Prince and Taylor, "Daniel Webster, the Boston Associates," 295.

55. Lowell and Appleton were more concerned with ensuring the safety of investors' capital and the size of their dividends than with developing long-range plans to increase production dramatically. Dalzell, *Enterprising Elite*, 55–56.

56. Dalzell, *Enterprising Elite*, 58.

57. John Quincy Adams to T. H. Perkins, Oct. 3, 1827, *Thomas Handasyd Perkins Papers*, MHS.

58. Richard Candee, "The Great Factory at Dover, New Hampshire: The Dover Manufacturing Co. Print Works, 1825," Old Sturbridge Village, MA; and Dalzell, *Enterprising Elite,* 49.

59. Dalzell, *Enterprising Elite,* 105.

60. Edward Balleisen argues that in antebellum America, certain people increasingly made a living off others' business failures. Edward J. Balleisen, "Vulture Capitalism in Antebellum America: the 1841 Federal Bankruptcy Act and the Exploitation of Financial Distress," *Business History Review* 70, no. 4 (1996): 474.

61. US Department of State, *Records of the 1820 Census of Manufactures, Schedule for Massachusetts and Rhode Island,* Microcopy number 279, National Archives, Washington, 1964, roll 2.

62. Jonathan Morrill and Son, Amesbury, MA, to Jeremiah Nelson, Washington, February 14, 1816, Volume 4, Records Collection of the Office of Naval Records and Library, Record Group 25, Entry 328, National Archives Building, Washington.

63. Candee, "The Great Factory," 39–41.

64. "To Sundry Accounts for shares held in the Capital stock of Cocheco Manufacturing Company October 14, 1827," Volume B-1 Stockholders' Records, Cocheco Manufacturing Company Records, BLHC.

65. "Miscellaneous" February 1829, Accounts Current, Volume 35, Boston Manufacturing Company Records, BLHC; and Gregory, *Nathan Appleton*, 152.

66. Entry, December 31, 1823, Vol. 2 Stockholders' Minutes, Merrimack Manufacturing Company Collection, BLHC.

67. Entry for January 24, 1826, Volume 3, Boston Manufacturing Company Collection, BLHC.

68. Theodore Steinberg argues the Boston Associates essentially created a "hydraulic empire" on the Merrimack River with these purchases of land and their control of "flowing lands." Theodore Steinberg, *Nature Incorporated: Industrialization and the Waters of New England* (New York: Cambridge University Press, 2004), 83; and "Miscellaneous," March 1822, Vol. 35, Boston Manufacturing Company Collection, BLHC.

69. John M. Cudd, *Chicopee Manufacturing Company, 1823–1915* (Lanham, MD: Rowman & Littlefield, 1974), 20.

70. Appleton, *Introduction of the Power Loom,* 30; entries October 23, 1821, and August 9, 1823, Volume 2 Directors' Records, Boston Manufacturing Company Collection, BLHC.

71. James Stimpson to Patrick Tracy Jackson, November 13, 1820, Box 6, Folder 9, Boston Manufacturing Company Collection, BLHC.

72. Moses Carleton to Patrick Tracy Jackson, December 17, 1820, Box 6, Folder 9, Boston Manufacturing Company Collection, BLHC.

73. Patent for Double Speeder, January 12, 1823, in the Aza Arnold Papers (Ms266), Box 2. RIHS.

74. ASP, "Protection to Manufactures," Finance 3, no. 568 (Jan. 10, 1820): 452.

75. Frank William Taussig, *Some Aspects of the Tariff Question: An Examination of the Development of American Industries under Protection*, vol. 12 (Cambridge: Harvard University Press, 1931); and Douglas A. Irwin and Peter Temin, "The Antebellum Tariff on Cotton Textiles Revisited," *Journal of Economic History* 61, no. 3 (Sept. 2001): 777–798. In a rejoinder, Harley faults Irwin's and Temin's regression model, which, he argues, incorrectly assumes that British and American factories produced different goods. Applying a Ricardian perception of trade, Harley asserts that Britain withheld cloth similar to American-made textiles from US markets, but not from, say, India or Brazil, precisely because the US tariff was prohibitive. C. Knick Harley, "The Antebellum Tariff: Different Products or Competing Sources? A Comment on Irwin and Temin," *Journal of Economic History* 61, no. 3 (Sept. 2001): 799–805.

76. Mathew Carey, *Address of the Philadelphia Society for the Promotion of National Industry, to the Citizens of the United States* (Philadelphia: n.p., 1819), 12.

77. Mathew Carey, *A View of the ruinous consequences of a dependence on*

foreign markets for the sale of the great staples of this nation, flour, cotton, and tobacco (Philadelphia: Mathew Carey, 1820), viii–ix.

78. N. R. Knight to Zachariah Allen, December 14, 1823, Box1, Folder 2, Zachariah Allen Papers, Ms254, RIHS.

79. Daniel Peart, "Looking beyond Parties and Elections: The Making of United States Tariff Policy during the Early 1820s," *Journal of the Early Republic* 33, no. 1 (Spring 2013): 87–108.

80. Irwin and Temin, "The Antebellum Tariff," 779.

81. *The American tariff, for 1833: comprising an alphabetical list of the duties on merchandize imported into the United States, in which the numerous errors and mistakes in the editions hitherto published are corrected* (London: O. Rich, 1833), 14. This rate, though, would be reduced by the compromise tariff of 1833.

82. William R. Bagnall, *The Textile Industries of the United States* (Cambridge, MA: Riverside Press, 1893), 359.

83. Stanwood, *American Tariff Controversies*, vol. 1, 207–210. See, for example, Account of Pawlet Manufacturing Co. Cloth Factory, Rutland County, US Department of State, *Records of the 1820 Census of Manufactures, Schedule for Vermont,* Microcopy number 279, National Archives, Washington, 1964, roll 3; "Answers of Milton Brown, of Pawlet," Document 7, Number 2. United States Department of the Treasury; and Louis McLane, *Documents Relative to the Manufactures in the United States*, vol. 1 (New York: Burt Franklin, 1969). The company was worth $17,000 at its establishment in 1809, and only $14,000 in 1832.

84. November 26, 1817, and February 18, 1819, Directors' Records, Volume 2, Boston Manufacturing Co. Collection. BLHC.

85. Carey, *A View of the ruinous consequences*, xiii.

86. David F. Ericson has argued that warfare and border patrol—hallmarks of European fiscal-military states and twentieth century America—characterized the US federal government's efforts to manage a slaveholding society. Ericson, *Slavery in the American Republic: Developing the Federal Government, 1791–1861* (Lawrence, University Press of Kansas, 2011).

87. Mark B. Wilson, "The Politics of Procurement: Military Origins of Bureaucratic Autonomy," *Journal of Policy History* 18, no. 1 (Jan. 2006): 45–75.

88. Jesse Slocumb, February 22, 1819, in Noble E. Cunningham Jr., ed., *Circular Letters of Congressmen to Their Constituents, 1789–1829* (Chapel Hill: University of North Carolina Press, 1978), 1055–1056.

89. $1,588,848 appropriated in 1837. ASP, "Estimates and Appropriations for Suppressing Hostilities of the Seminole Indians in Florida," 25th Congress, 1st Session, no. 739 (Sept. 15, 1837): 466. By 1840, $11,574,477 had been spent. ASP, "Expenditures in Suppressing Indian Hostilities in Florida," 26th Congress, 2nd Session, no. 8 (Dec. 15, 1840): 10.

90. Stagg, *Mr. Madison's War*, 504. For the roots of American anti-militarism, see Richard H. Kohn, *Eagle and Sword: The Federalists and the Creation of the Military Establishment in America, 1783–1802* (New York: Free Press, 1975), 3. Kohn argues that the origins of Americans' fear of armies lay in colonial tax and radical Whig ideology.

91. Noble E. Cunningham Jr., ed., *Circular Letters of Congressmen to Their Constituents, 1789–1829* (Chapel Hill: University of North Carolina Press, 1978); and ASP, "Reduction of the Army Considered," Military Affairs 1, no. 168 (Dec. 4, 1818): 779.

92. Beginning in 1835, the second round of the Florida Seminole Wars absorbed a tremendous amount of resources (between $30 million and $40 million—50 percent of annual expenditures—and forty thousand troops). ASP, "Estimates and Appropriations for Suppressing Hostilities of the Seminole Indians in Florida," 25th Congress, 1st Session, no. 739 (Sept. 15, 1837): 466; and ASP, "Expenditures in Suppressing Indian Hostilities in Florida," 26th Congress, 2nd Session, no. 8 (Dec. 15, 1840): 10. Annual federal expenditures, exclusive of public debt, were $17.5 million in 1835, $30.9 million in 1836, $39.2 million in 1837. ASP, "Statement showing the amount of annual estimates submitted by the Secretary of Treasury, and annual expenditures," 25th Congress, 2nd Session, no. 497 (June 29, 1838): 16–20.

93. Historian William H. Bergmann has demonstrated for the early national period the importance of military bureaucracy for managing Indian conflicts and controlling supply lines to western territories. Bergmann, *The American National State and the Early West* (New York: Cambridge University Press, 2012), 9. For the importance of controlling supply lines, see chapter two. For the dominant role bureaucracy has played in driving economic change/business history in the United States, see Charles Perrow, *Organizing America: Wealth, Power, and the Origins of Corporate Capitalism* (Princeton: Princeton University Press, 2009).

94. ASP, "Extension of the Ordnance Department," Military Affairs 1, no. 121 (June 19, 1813): 336.

95. "An Act for the better regulation of the Ordnance Department," February 8, 1815, 13th Congress, 3rd Session, ch. 38, in Richard Peters, *United States Statutes at Large,* vol. 3 (GPO, 1846), 203–204.

96. Secretary of War John C. Calhoun, for example, argued that arms could be fabricated better and at least as cheaply by the federal armories as by contract. ASP, "Western Armory," Military Affairs 1, no. 166 (Dec. 7, 1818): 773; Merritt Roe Smith, *Military Enterprise and Technological Change: Perspectives on the American Experience* (London: MIT Press, 1985), 60; and Felicia Johnson Deyrup, *Arms Makers of the Connecticut Valley: A Regional Study of the Economic Development of the Small Arms Industry, 1798–1870* (Northampton, MA, 1948), 55.

97. Rodrigo Canales, "From Ideals to Institutions: Institutional Entrepreneurship and the Growth of Mexican Small Business Finance," Yale School of Management, working paper (2011), 3; Katherine C. Kellogg, "Operating Room: Relational Spaces and Microinstitutional Change in Surgery," *American Journal of Sociology* 115, no. 3 (Nov. 2009): 657–711; Heather A. Haveman and Hayagreeva Rao, "Structuring a Theory of Moral Sentiments: Institutional and Organizational Coevolution in the Early Thrift Industry," *American Journal of Sociology* 102, no. 6 (May 1997): 1606–1651; and Pamela S. Tolbert and Lynne G. Zucker, "Institutional Sources of Change in the Formal Structure of Organizations: The Diffusion of Civil Service Reform, 1880–1935," *Administrative Science Quarterly* (1983): 22–39.

98. The Springfield Armory complied successfully with this order; Harpers Ferry did not. Smith, *Military Enterprise and Technological Change,* 53.

99. Joshua L. Rosenbloom, "Anglo-American Technological Differences in Small Arms Manufacturing," *The Journal of Interdisciplinary History* 23, no. 4 (Spring 1993): 691. Also, US troops used rifles and muskets, which were less precise and therefore could more easily be made by interchangeable manufacture. Robert A. Howard, "Interchangeable Parts Reexamined: The Private Sector of the American Arms Industry on the Eve of the Civil War," *Technology and Culture* 19, no. 4 (Oct. 1978): 649.

100. US gun makers produced the first fully interchangeable firearm in 1841. Smith, *Military Enterprise and Technological Change,* 61–63.

101. Decius Wadsworth to Roswell Lee, July 13, 1817, Box 2, Folder 8, SA-LRM.

102. Weeks, *John Quincy Adams and American Global Empire,* 107.

103. Weeks, *John Quincy Adams and American Global Empire*, 109–110; and ASP, "War with the Seminole Indians," Military Affairs 1, no. 163 (March 25, 1818): 680.

104. ASP, "Reduction of the Army Considered," Military Affairs 1, no. 168 (Dec. 4, 1818): 779; ASP, "Reduction of the Army," 15th Congress, 2nd Session (Dec. 11, 1818): 391.

105. Decius Wadsworth to Roswell Lee, January 27, 1817, Box 2, Folder 4, SA-LRM.

106. John Morton to Roswell Lee, December 16, 1817, Box 3, Folder 3, SA-LRM.

107. Decius Wadsworth to Roswell Lee, November 10, 1818, Box 1, Target #2, SA-LRO.

108. Decius Wadsworth to Roswell Lee, July 25, 1818, Box 1, Target #2, SA-LRO.

109. Decius Wadsworth to Roswell Lee, Springfield Armory, August 30, 1817, Box 3, Folder 9, SA-LRM.

110. Smith, *Military Enterprise and Technological Change*, 57.

111. John Morton to Roswell Lee, August 30, 1817, Box 3, Folder 9, SA-LRM.

112. Adam Kinsley to Roswell Lee, April 7, 1817, Box 2, Folder 6, SA-LRM.

113. David R. Meyer, *Networked Machinists: High-Technology Industries in Antebellum America* (Baltimore: Johns Hopkins University Press, 2006), 231.

114. "Wool Carding," [Springfield] *Hampden Federalist*, June 19, 1817; and Benjamin Jenks and George Wilkinson to Roswell Lee, April 19, 1817, Box 3, Folder 6, SA-LRM.

115. Merritt Roe Smith, *Harpers Ferry Armory and the New Technology: The Challenge of Change* (Ithaca: Cornell University Press, 1977), 325–326.

116. Decius Wadsworth to William H. Crawford, April 20, 1816, quoted in Simeon Newton Dexter North, *Simeon North, First Official Pistol Makers of the United States*; and *A Memoir* (Concord, NH: Rumford Press), 103–105.

117. Merritt Roe Smith, "John H. Hall, Simeon North, and the Milling Machine: The Nature of Innovation among Antebellum Arms Makers," *Technology and Culture* 14, no. 4 (Oct. 1973): 576.

118. Smith, "John H. Hall, Simeon North, and the Milling Machine," 574.

119. Decius Wadsworth to Roswell Lee, March 8, 1816, Box 1, Letters Received Miscellaneous, Record Group 156, Entry 1362, NM-59, 94-066, National Archives Building, Waltham, MA; and Adonijah Foot to Roswell Lee, December 7, 1824, Box 11, Folder 1, SA-LRM.

120. Simeon North to Colonel George Talcott, May 15, 1839, in James E. Hicks, *United States Ordnance: Volume II, Ordnance Correspondence Relative to Muskets, Rifles, Pistols and Swords* (New York: James E. Hicks, 1940), 114. The rifle was, in fact, novel enough that it required an instruction manual. "Directions for Using the Arms known by the name Hall's Rifle," Box 1, Folder 10, Firearm Makers Collection, MCHS.

121. In fact, Jacob Perkins, inventor of the steam gun, left the United States for Great Britain in the 1820s for this reason. Greville Bathe and Dorothy Bathe, *Jacob Perkins: His Inventions, His Times, & His Contemporaries* (Philadelphia: Historical Society of Pennsylvania, 1943), 109–111; and Jerome V. C. Smith, ed., *Bowen's Boston News-letter, and City Record* (Boston: A. Bowen, 1826), 350.

122. Decius Wadsworth to Roswell Lee, February 6, 1819, Box 1, Target #3, SA-LRO.

123. Decius Wadsworth to Roswell Lee, February 22, 1816, Box 1, Folder 5; and Decius Wadsworth to Roswell Lee, February 22, 1816, Box 1, Folder 5, SA-LRM.

124. Nathan Starr Sr. to Nathan Starr Jr., March 24, 1818, Folder 1, Box 1, Arms Correspondence, Starr Family Collection, MCHS.

125. Nathan Starr Sr. to Nathan Starr Jr., March 25, 1818, Folder 1, Box 1, Arms Correspondence, Starr Family Collection, MCHS.

126. Robert Johnson to Roswell Lee, December 25, 1817, Box 2, Folder 3, SA-LRM; and Robert Johnson to Roswell Lee, February 27, 1819, Box 4, Folder 6, SA-LRM.

127. George Bomford to Roswell Lee, March 6, 1823, Box 2, SA-LRO.

128. ASP, "Contracts Made since January 1, 1820 for canon, shot, muskets, and other small arms," Military Affairs 2, no. 248 (Jan. 6, 1824): 599.

129. "An Act to Reduce and Fix the Military Peace Establishment of the United States," 16th Congress, 2nd Session, 23 (March 2, 1821): 615.

130. ASP, "Ordnance Department," 18th Congress, 1st Session (Dec. 30, 1823): 905–906.

131. George Bomford to Roswell Lee, March 12, 1824, Box 2, SA-LRO.

132. George Bomford to Roswell Lee, August 18, 1825, Box 2, SA-LRO.

133. William Wade to Roswell Lee [no date], Box 2, SA-LRO.

134. George Bomford to Roswell Lee, August 18, 1825, Box 2, SA-LRO; and William Wade to Roswell Lee November 30, 1828, Letterbook Volume 1, SA-LRO.

135. ASP, "Statement of the Expenditures of The United States Armory at Springfield, Mass., and of the arms made and repaired therein, from its establishment to the close of the year 1821," 17th Congress, 2nd Session, no. 111 (Nov. 30, 1822): 84.

136. ASP, Register of Debates, 20th Congress, 2nd Session (Dec. 17, 1828): 112.

137. Roswell Lee to Adonijah Foot, August 6, 1823, Folder 5, Box 9, SA-LRM.

138. William Wade to Roswell Lee, November 30, 1828, Letterbook Volume 1, SA-LRO.

139. ASP, "Ordnance Department," 18th Congress, 1st Session (Dec. 1823): 905–906; Journal of the Senate, Vol. 13, 18th Congress, 1st Session (Dec. 2, 1823): 14; "Ordnance Bill," Register of Debates, House of Representatives, 22nd Congress, 1st Session (March 22, 1832): 2239; Register of Debates, House of Representatives, 20th Congress, 2nd Session (Dec. 17, 1828): 112; "An Act to Reduce and Fix the Military Peace Establishment of the United States," March 2, 1821, in Trueman T. Cross, *Military Laws of the United States* (Washington: Edward de Krafft, 1838), 213; "An Act Providing for the Organization of the Ordnance Department," 22nd Congress, 1st Session, 67 (April 5, 1832): 504; and ASP, "Ordnance Bill," 22nd Congress, 1st Session (March 22, 1832): 2237–2238. For Seminole opposition to treaties with the US government, see Andrew Welch, *A narrative of the early days and remembrances of Oceola Nikkanochee, prince of Econchatti* (London: Hatchard, 1841), 212–215.

140. US Congress, *Reports of Committees of the House of Representatives* (Washington, 1973), 125–126.

141. For the most recent scholarship on the Second Seminole War, see C. S.

Monaco, *The Second Seminole War and the Limits of American Aggression* (Baltimore: Johns Hopkins University Press, 2018). For an account of Seminoles' persistent resistance to removal, made possible by their effective system of arms supply, see David Silverman, *Thundersticks: Firearms and the Violent Transformation of Native America* (Cambridge: Belknap Press of Harvard University Press, 2016), 190–220.

142. Colin G. Calloway, *Pen and Ink Witchcraft: Treaties and Treaty Making in American Indian History* (New York: Oxford University Press, 2013).

143. ASP, "Seminole Hostilities," 24th Congress, 1st Session, no. 271 (June 3, 1836): 246–247.

144. M. L. Brown, "Notes on U.S. Arsenals, Depots, and Martial Firearms of the Second Seminole War," *Florida Historical Quarterly* 61, no. 4 (Apr. 1983): 447.

145. ASP, "Seminole Hostilities," 24th Congress, 1st Session, no. 271 (June 3, 1836): 246, 194.

146. For a discussion of Seminole arming strategy, see Silverman, *Thundersticks*, 200.

147. ASP, "Hostile Indians in Florida," 24th Congress, 1st Session, no. 38 (Jan. 5, 1836).

148. "Documents Accompanying the Report of the Secretary of War," 1840, 50; and ASP, "Expenditures in Suppressing Indian Hostilities in Florida," 26th Congress, 2nd Session, no. 8 (Dec. 15, 1840): 6.

149. "Bond of the Officers of the Tallahassee Volunteers," February 10, 1840, Volume 2, Contracts for Ordnance and Ordnance Supplies, Records of the Chief of the Ordnance Department, Record Group 156, Entry 78, National Archives, Washington.

150. Andrew Jackson to Joel Roberts Poinsett, August 27, 1837, Folder 3, Box 9, Joel Roberts Poinsett Papers (Collection 0512), Historical Society of Pennsylvania (hereafter JRPP).

151. James Gadsden to Joel Roberts Poinsett, September 17, 1837, Folder 6, Box 9, JRPP.

152. Andrew Jackson to Joel Roberts Poinsett, August 27, 1837, Folder 3, Box 9, JRPP.

153. James Gadsden to Joel Roberts Poinsett, August 24, 1837, Folder 3, Box 9, JRPP.

154. For a description of the nature of warfare in Florida, see John T. Sprague, *The Origin, Progress, and Conclusions of the Florida War* (New York: Appleton, 1848).

155. Andrew Jackson to Joel Roberts Poinsett, October 14, 1837, Folder 8, Box 9, JRPP. Brown, "Notes on U.S. Arsenals, Depots, and Martial Firearms," 450.

156. ASP, "Report of the President of a Board of Officers on Improvements

in Fire-Arms by Hall, Colt, Cochran, Leavitt, and Baron Hackett, as Compared with the United States Musket," Military Affairs 7, no. 743 (Oct. 3, 1837): 526.

157. [Tallahassee] *The Floridian*, January 13, 1838, 1.

158. Andrews, *Communication from the Secretary of the Treasury*, 772. Over time, however, violence would begin to erase Florida's frontier past and it would take its place among the "Old South." Edward Baptist, *Creating an Old South: Middle Florida's Plantation Frontier before the Civil War* (Chapel Hill: University of North Carolina Press, 2002).

159. Douglass North, *The Economic Growth of the United States, 1790–1860* (Englewood Cliffs, NJ: Prentice-Hall, 1961), 156; and Stanwood, *American Tariff Controversies*, vol. 1, 160–162.

160. The United States spent $30 million on the Second Seminole War. David F. Ericson, *Slavery in the American Republic: Developing the Federal Government* (Lawrence: University Press of Kansas, 2011), 16. Adams and Monroe chose not to negotiate for Texas because they did not want it. Onís had already agreed to cede Florida at the beginning of 1818; the signing of the treaty was delayed another year not because Adams tried to acquire Texas, but because he struggled to gain as much of the Northwest as possible. Weeks, *John Quincy Adams and American Global Empire,* 167.

161. ASP, "List of Documents Accompanying the Report of the Secretary of War," no.18 (Feb. 13, 1851): 154.

162. The Springfield standard pattern was firmly in place by 1823. Smith, *Military Enterprise and Technological Change,* 60. Samuel Lathrop, US Representative from Massachusetts, dined with President Monroe, who informed him that he was convinced of the importance of the Springfield Armory and very pleased with the work it had done. Samuel Lathrop to Roswell Lee, January 1, 1820, Box 5, Folder 7, SA-LRM.

163. Years of government contracts, for example, had put Simeon North's business on sound footing, which enabled him to transition from pistol to rifle manufacture, for which he received a lucrative contract in 1824. North, *Simeon North, First Official Pistol Makers*, 156–157.

164. See, for example, Tench Coxe, Purveyor of Public Supplies, Philadelphia, to the Manufacturers of the United States, July 17, 1810, Unbound Papers 1801–1811, Almy and Brown Records Ms29, RIHS.

165. ASP, "Clothing the Army with Domestic Fabrics," Military Affairs 2, no. 180 (Jan. 11, 1820): 42.

Chapter 5. Managing New Markets

1. William Tudor to Henry Clay, June 12, 1825, US Department of State, Despatches from US consuls, RG59, National Archives, Washington (hereafter USDS-DC), Lima.

2. William Tudor to Henry Clay, June 12, 1825, USDS-DC Lima.

3. After speaking to Marshal Santa Cruz, the council of government deliberated on Tudor's recommendation and decided to lessen the duty. William Tudor to Henry Clay, November 21, 1826, USDS-DC Lima.

4. William Tudor to Henry Clay, February 28, 1826, USDS-DC Lima.

5. Merchants themselves did not necessarily even want free trade. See, for example, a letter to Rhode Island merchant Edwin T. Jenckes from his business associates in Arica, Chile, in which they express the hope that the Chilean government will levy discriminatory tariffs that would enable their cargoes to compete more effectively against the glut of British goods. McFadon and Cobb to Edwin T. Jenckes, July 26, 1825, Box 8, Folder 1, Nightingale-Jenckes Papers (MSS588), RIHS.

6. For the role of early US commercial diplomacy in other regions of the world, see, for example, David W. McFadden, "John Quincy Adams, American Commercial Diplomacy, and Russia, 1809–1825," *New England Quarterly* 66, no. 4 (Dec. 1993): 613–629, and Ruth Kark, *American Consuls in the Holy Land, 1832–1914* (Detroit, MI: Wayne State University Press, 1994). For the importance of merchants as consuls and diplomats to US commerce, see John M. Belohlavek, "Economic Interest Groups and the Formation of Foreign Policy in the Early Republic," *Journal of the Early Republic* 14, no. 4 (Winter 1994): 479.

7. In terms of commercial diplomacy during the wars, this chapter is focused mainly on Spanish South America. Brazil experienced relatively more stability during the 1810s because Portugal moved its capital there during the Napoleonic Wars. Americans did, however, generally support Brazilian independence in the 1820s, especially because it was one of the few Latin American nations to maintain the institution of slavery. As Caitlin Fitz has shown, US citizens and policy makers viewed their relationship with Mexico differently than the rest of South America because its proximity offered both more opportunities and threats for US territorial ambitions. Caitlin A. Fitz, *Our Sister Republics: The United States in an Age of American Revolution* (New York: W. W. Norton, 2016). Although the United States' relationships with Latin American colonies and nations varied, in general US diplomats by the 1820s demonstrated increasing aggression in their pursuit of US economic interests throughout the region.

8. For the impact of imported arms on the patriot cause, see Rafe Blaufarb, "Arms for Revolutions: Military Demobilization after the Napoleonic Wars and Latin American Independence," in Alan Forrest, Karen Hageman, and Michael Rowe, eds., *War, Demobilization and Memory: The Legacy of War in the Era of Atlantic Revolutions* (Basingstoke, UK: Palgrave Macmillan, 2016), 100–116. For American re-exports of European arms to South America, see Caitlin A. Fitz, "Our Sister Republics: The United States in an Age of American Revolution" (PhD diss., Yale University, 2010), appendix 3.

9. While Mary Rose maintains that US manufacturers depended more on domestic tariff support than their counterparts in Britain who traded freely in

the international marketplace, she downplays the extent to which export markets became increasingly important for the United States. Mary B. Rose, *Firms, Networks, and Business Values: The British and American Cotton Industries since 1750* (Cambridge: Cambridge University Press, 2000).

10. Brian DeLay, "How Not to Arm a State: American Guns and the Crisis of Governance in Mexico, Nineteenth and Twenty-First Centuries," *Southern California Quarterly* 95, no. 1 (Spring 2013): 8–9. For surplus stock see John Morton, Ordnance Department, to Roswell Lee, Superintendent, Springfield Armory, July14, 1817, Box 2, Folder 8, SA-LRM.

11. Soft power can be understood as the upper hand gained by engaging in economic relationships that involve discrimination. Jonathan Kirshner, "Political Economy in Security Studies after the Cold War," *Review of International Political Economy* 5, no. 1 (Spring 1998): 72. For twentieth-century "dollar diplomacy" in Latin America, see David Sheinin, "The New Dollar Diplomacy in Latin America," *American Studies International* 37, no. 3 (October 1999): 81–99.

12. For a discussion of the importance of international trade for the US economy, and the subsequent shift to manufacturing investment, see chapter five of Douglass Cecil North, *The Economic Growth of the United States, 1790–1860* (Englewood Cliffs, NJ: Prentice-Hall, 1961), 46–58. For Atlantic commerce during the Age of Revolutions, see James F. Shepherd and Gary M. Walton, "Economic Change after the American Revolution: Pre- and Post-War Comparisons of Maritime Shipping and Trade," *Explorations in Economic History* 13, no. 4 (October 1976): 397–422; Claudia D. Goldin and Frank D. Lewis, "The Role of Exports in American Economic Growth during the Napoleonic Wars, 1793 to 1807," *Explorations in Economic History* 17, no. 1 (January 1980): 6–25; Donald R. Adams Jr., "American Neutrality and Prosperity, 1793–1808: A Reconsideration," *Journal of Economic History* 40, no. 4 (December 1980): 71; and Brooke Hunter, "Wheat, War, and the American Economy during the Age of Revolution," *William and Mary Quarterly*, 3rd Series 62, no. 3 (July 2005): 505–526, 983. For the American Atlantic trade more broadly, see John H. Coatsworth, "American Trade with European Colonies in the Caribbean and South America, 1790–1812," *William and Mary Quarterly* 24, no. 2 (Apr. 1967): 243–266; Javier Cuenca Esteban, "Trends and Cycles in US Trade with Spain and the Spanish Empire, 1790–1819," *Journal of Economic History* 44, no. 2 (June 1984): 521–543; Javier Cuenca Esteban, "British 'Ghost' Exports, American Middlemen, and the Trade to Spanish America, 1790–1819: A Speculative Reconstruction," *William and Mary Quarterly* 71, no. 1 (Jan. 2014): 63–98; Michelle Craig MacDonald, "The Chance of the Moment: Coffee and the New West Indies Commodities Trade," *William and Mary Quarterly*, 3rd Series 62, no. 3 (July 2005): 441–472; Linda Kerrigan Salvucci, "Development and Decline: The Port of Philadelphia and

Spanish Imperial Markets, 1783–1823" (PhD diss., Princeton University, 1985); Jacques A. Barbier and Allan J. Kuethe, eds., *The North American Role in the Spanish Imperial Economy, 1760–1819* (Manchester, UK: Manchester University Press, 1984); Edward P. Pompeian, "Spirited Enterprises: Venezuela, the United States, and the Independence of Spanish America, 1789–1823" (PhD diss., The College of William and Mary, 2014); Esteban, "British 'Ghost' Exports," 63–98; and James Alexander Dun, "'What Avenues of Commerce, Will You, Americans, Not Explore!': Commercial Philadelphia's Vantage onto the Early Haitian Revolution," *William and Mary Quarterly* 62, no. 3 (July 2005): 473–504.

13. See responses to "the different kinds and values of foreign materials" and "what proportion in foreign countries and where," in United States Department of the Treasury and Louis McLane, *Documents Relative to the Manufactures in the United States,* vol. 1 (New York: Burt Franklin, 1969). And these are only the ones who reported it. Many sent goods to Philadelphia, New York, and Baltimore, and from there agents shipped their goods to South America and Mexico. For example, W. A. Andross, the agent for Hartford Cotton Manufacturing Company, responded that his company's goods were sold in the United States and "if exported, not known." United States Department of the Treasury and McLane, *Documents Relative to the Manufactures,* vol. 1, 1006. There were 249 firms capitalized at over $50,000, but the majority were much smaller. Alfred D. Chandler, *The Visible Hand: The Managerial Revolution in American Business* (Cambridge: Belknap Press of Harvard University Press, 1977), 60–61.

14. While the United States' biggest export was cotton to Great Britain, the United States sent mostly manufactured goods to Latin America. ASP, "Exports for the Year ending in September 30 1815," Commerce and Navigation 15, no. 196 (Feb. 15, 1816): 20; "Exports for the Year Ending September 30, 1816," Commerce and Navigation 2, no. 205 (Feb. 3, 1817): 55; "Commerce and Navigation for Year Ending September 30, 1821," Commerce and Navigation 2, no. 246 (Jan. 24, 1822): 239; US House, Committee on Commerce and Navigation, "Letter from the Secretary of the Treasury, in relation to the commerce and navigation of the United States for the year 1836," 25th Congress, 1st Session, no. 188 (April 4, 1837): 731; see also United States Bureau of the Census, *Historical Statistics of the United States, Colonial Times to 1970* (Washington: US Dept. of Commerce, Bureau of the Census, 1976), 551.

15. General Convention of Peace, Amity, Navigation, and Commerce, United States and the Republic of Colombia, 1824. Convention of Peace, Amity, Commerce, and Navigation, United States and Republic of Chile, May 16, 1832, proclaimed April 29, 1834. Treaty of Peace, Friendship, Navigation, and Commerce, United States and the Republic of Venezuela, signed on January 20, 1836. Treaty of Peace, Friendship, Navigation, and Commerce, United States and Ecuador, June 13, 1839.

16. Alexander Hamilton, *Report on Manufactures*, December 5, 1791, in Harold C. Syrett, ed., *The Papers of Alexander Hamilton,* vol. 19 (New York: Columbia University Press, 1973).

17. For the maintenance of neutrality, see William G. Miller to John Quincy Adams, September 14, 1821, USDS-DC Montevideo. Miller remarked to John Quincy Adams in regard to his dealing with Portuguese officials during their occupation of Montevideo: "I have studiously avoided any compromise that would hereafter render me as consul of the United States unpopular with the patriot party." For examples of requests for military and political support see Thomas Lloyd Halsey to James Monroe, May 5, 1815, USDS-DC Buenos Aires, and William Worthington to John Quincy Adams, January 1, 1818, USDS, Despatches from US Ministers to Argentina, 1817–1906, RG59, National Archives, Washington.

18. Oliver Pollock to Robert Livingston, December 14, 1783, USDS-DC Havana.

19. John Morton to James Madison, January 20, 1802, USDS-DC Havana.

20. Bibiano Torres Ramirez and Javier Ortiz de la Tabla, eds., *Reglamento para el Comercio Libre, 1778* (Seville, 1979); and John Fisher, "Imperial 'Free Trade' and the Hispanic Economy, 1778–1796," *Journal of Latin American Studies* 13, no. 1 (May 1981): 21. Venezuela and New Spain were excluded from this "free trade" between Spain and the Americas. The Caracas Company had successfully dominated trade between Caracas and Vera Cruz and it was feared that if this trade were opened to other Spanish merchants, they would all flock there to the detriment of other colonies' trade. For an explanation of the ways in which the New World silver trade enabled Spain to maintain a tobacco monopoly, see Carlos Marichal and Matilde Souto Mantecón, "Silver and Situados: New Spain and the Financing of the Spanish Empire in the Caribbean in the Eighteenth Century," *Hispanic American Historical Review* 74, no. 4 (Nov. 1994): 602.

21. James R. Fichter, *So Great a Profit: How the East Indies Trade Transformed Anglo-American Capitalism* (Cambridge: Harvard University Press, 2010).

22. James Ferguson King, "The Evolution of the Free Trade Principle in Spanish Colonial Administration, *Hispanic American Historical Review* 22, no. 1 (Feb. 1942): 34–56, 55–56. See also Leonardo Marques, "Slave Trading in a New World: The Strategies of North American Slave Traders in the Age of Abolition," *Journal of the Early Republic* 32, no. 2 (Summer 2012): 233–260, for a discussion of changing attitudes toward the slave trade, as well as a discussion of the importance of Cuba as a market for American slave traders.

23. William G. D. Worthington to Richard Rush, April 28, 1817, USDS, Despatches from US Ministers to Argentina, 1817–1906, RG59, National Archives, Washington: 1952. Microfilm, no. M-69, roll 1.

24. See, for example, P. Sartoris to José Bonifacio de Silva, August 21, 1822, USDS-DC Rio de Janeiro.

25. Charles Stuart Kennedy, *The American Consul: A History of the United States Consular Service, 1776–1914* (New York: Greenwood, 1990), 20.

26. Kennedy, *The American Consul*, 18.

27. Kennedy, *The American Consul*, 21.

28. See, for example, John Quincy Adams to James Birkhead, and John Quincy Adams to Aaron Tyler, August 19, 1819, USDS, Domestic Letters of the Department of State, RG59, National Archives, Washington.

29. Benjamin Franklin, John Adams, and Arthur Lee, joint letter to President of Congress, July 20, 1778, in Charles Francis Adams, *The Works of John Adams, with a Life of the Author* (Boston: Little, Brown, 1856), 20.

30. John Morton to James Madison, December 11, 1801, USDS-DC Havana.

31. Ellen Hartigan O'Connor, "'Auctioneer of Offices': Patronage, Value, and Trust in the Early Republic Marketplace," *Journal of the Early Republic* 33, no. 3 (Fall 2013): 463–488.

32. Kennedy, *The American Consul*, 52; United States Department of State, *General Instructions to the Consuls and Commercial Agents of the United States* (Washington, 1838), 24.

33. See, for example, John M. Forbes to John Quincy Adams, December 29, 1821, USDS-DC Buenos Aires; John Quincy Adams to Messrs. Peabody and Tucker, February 10, 1818, USDS, Domestic Letters of the Department of State, RG59, National Archives, Washington.

34. See, for example, Simeon Baldwin and Francis J. Spooner to John MacPherson, May 17, 1824, and Simeon Baldwin and Francis J. Spooner to Prescott and Sherman, May 6, 1824, Baldwin and Company Papers (MS 56), Yale University Manuscripts and Archives, Box 1, Letterbook 1823–1824. New York shipping firm Baldwin and Spooner depended on US consul to Cartagena John MacPherson to help them pursue their claims against the Colombian government.

35. Gervasio Antonio de Posadas to James Madison, March 9, 1814, USDS-DC Buenos Aires. For another example, see Thomas Lloyd Halsey to James Monroe, May 3, 1813, USDS-DC Buenos Aires, in which Halsey informs Monroe that the government of Buenos Aires has expressed its desire for a large quantity of military supplies.

36. Robert K. Lowry to Robert Smith, October 1, 1810, USDS-DC La Guiara.

37. See, for example, William G. Miller to James Monroe, March 15, 1812, USDS-DC Montevideo, and Thomas Lloyd Halsey to James Monroe, May 3, 1812, USDS-DC Buenos Aires.

38. See, for example, newspaper articles such as "From the Saratoga Journal," [New York] *Shamrock*, January 21, 1815, which deplores the "slavery and

oppression" of Old Spain in Spanish America and provides an overview of geography and the independence struggles to its readers, and "United Provinces of La Plata," [Charles Town, West Virginia] *Farmer's Repository*, February 29, 1816. For an excellent analysis of US support for Latin American independence movements, see Fitz, *Our Sister Republics*.

39. Robert K. Lowry to Robert Smith, October 1, 1810, and Lowry to Monroe, USDS-DC La Guiara.

40. Elton Atwater describes the law of neutrality as a working set of agreements between nations regarding behavior toward belligerents, in *American Regulation of Arms Exports* (Washington: Carnegie Endowment for International Peace, 1941), 7n2. John Missall and Mary Lou Missall, *The Seminole Wars: America's Longest Indian Conflict* (Gainesville: University Press of Florida, 2004), 52.

41. In a note to British Minister at Washington George Hammond on May 15, 1793, Jefferson announced that the government would not prohibit the private sale of arms or contraband to belligerents. Atwater, *American Regulation of Arms Exports*, 8.

42. Theodore Draper, *A Very Thin Line: The Iran-Contra Affairs* (New York: Hill and Wang, 1991); Keith Krause, *Arms and the State: Patterns of Military Production and Trade* (New York: Cambridge University Press, 1995); Jonathan A. Grant, *Rulers, Guns, and Money: The Global Arms Trade in the Age of Imperialism* (Cambridge: Harvard University Press, 2007); Joseph Bradley, *Guns for the Tsar: American Technology and the Small Arms Industry in Nineteenth-Century Russia* (Dekalb: Northern Illinois University Press, 1990); Katherine C. Epstein, *Torpedo: Inventing the Military-Industrial Complex in the United States and Great Britain* (Cambridge: Harvard University Press, 2014); Robert Higgs, ed., *Arms, Politics, and the Economy* (Oakland, CA: Holmes and Meier, 1990); Thomas Mahnken, Joseph Maiolo, and David Stevenson, eds., *Arms Races in International Politics* (New York: Oxford University Press, 2016); Brian DeLay, "How Not to Arm a State: American Guns and the Crisis of Governance in Mexico, Nineteenth and Twenty-First Centuries," *Southern California Quarterly* 95, no. 1 (Spring 2013): 5–23; Emrys Chew, *Arming the Periphery: The Arms Trade in the Indian Ocean During the Age of Global Empire* (Basingstoke, UK: Palgrave Macmillan, 2012); Andrew T. H. Tan, ed., *The Global Arms Trade: A Handbook* (Abingdon, UK: Routledge, 2014); Edward A. Kolodziej, *Making and Marketing Arms: The French Experience and its Implications for the International System* (Princeton: Princeton University Press, 2014); Ken Conca, *Manufacturing Insecurity: The Rise and Fall of Brazil's Military-Industrial Complex* (Boulder, CO: L. Rienner, 1997); and Mark Harrison, ed., *Guns and Rubles: The Defense Industry in the Stalinist State* (New Haven: Yale University Press, 2008).

43. Deyrup, *Arms Makers of the Connecticut Valley,* 46.

44. Nathan Starr, Middletown, CT, to Roswell Lee, Superintendent, Springfield Armory, Washington, January 30, 1826. Box 12, Folder 1, SA-LRM.

45. Auction houses were often viewed as the best place for the armory to dispose of its goods. John Morton, Ordnance Department, to Roswell Lee, Superintendent, Springfield Armory, July 5, 1815. Box 1, Folder 3, SA-LRM.

46. Martin Tovar Ponte to Robert Smith, April 25, 1810, USDS, *Notes from the Colombian Legation, 1810–1906*, RG59 National Archives, Washington. Soon after fighting broke out in Venezuela, Ponte wrote to Smith that because Spain was now occupied by a foreign monarch and a sovereign junta had been established for the territory of Venezuela, Venezuela would no longer accept goods from Spain's "mercantile expeditions" and would "await [US trade] with open arms."

47. Thomas Lloyd Halsey to James Madison, February 11, 1815, USDS-DC Buenos Aires.

48. William G. Miller to James Monroe, March 15, 1812, USDS-DC Montevideo; and Thomas Lloyd Halsey to James Monroe, May 3, 1812, USDS-DC Buenos Aires.

49. Thomas Lloyd Halsey to James Monroe, October 21, 1812, USDS-DC Buenos Aires; and Gervasio Antonio de Posadas to James Madison, March 9, 1814, USDS-DC Buenos Aires.

50. Early letters from the US consul in Havana complained of the "jealousies that appear by the officers of government against our trade" and the expectations that US trade with the island would "dwindle away to nothing." Oliver Pollock to Robert Livingston, December 14, 1783, USDS-DC Havana. The Spanish Crown believed in a "free," Crown-protected trade between European and American Spaniards, which operated to the detriment of American traders seeking outlets for European and Asian goods, as well as for American produce like tobacco. Ramirez and de la Tabla, eds., *Reglamento para el Comercio Libre*; and John Fisher, "Imperial 'Free Trade' and the Hispanic Economy, 1778–1796," *Journal of Latin American Studies* 13, no. 1 (May 1981): 21. The only sector in which Spain had encouraged truly "free" trade was slave trading, which Americans certainly took advantage of, both legally and illegally, but they soon needed to find other commodities to trade in South America, as many of Spain's former colonies enacted legislation that restricted slavery and the slave trade. James Ferguson King, "The Evolution of the Free Trade Principle in Spanish Colonial Administration," *Hispanic American Historical Review* 22, no. 1 (Feb. 1942): 55–56. See also Leonardo Marques, "Slave Trading in a New World: The Strategies of North American Slave Traders in the Age of Abolition," *Journal of the Early Republic* 32, no. 2 (Summer 2012): 233–260, for a discussion of changing attitudes toward the slave trade, as well as the importance of Cuba

as a market for American slave traders. And for the importance of silver for the development of American capitalism and the US financial sector, see Fichter, *So Great a Profit*, 31–33.

51. Decius Wadsworth to Roswell Lee, September 26, 1817, Box 3, Folder 1, SA-LRM.

52. Henry W. Huntington to Roswell Lee, September 20, 1818, Box 4, Folder 1, SA-LRM.

53. John Morton, Captain of Ordnance, to Roswell Lee, Superintendent, Springfield Armory, and William Cramond to Roswell Lee, Superintendent, Springfield Armory, June 19, 1815, and September 25, 1815, Box 1, Folders 3 and 4, SA-LRM; For the variation in bank notes' values, see Stephen Mihm, *Nation of Counterfeiters: Capitalists, Con Men, and the Making of the United States* (Cambridge: Harvard University Press, 2007).

54. William Cramond, Philadelphia, to Roswell Lee, Superintendent, Springfield Armory, July 17, 1815. Box 1, Folder 9, SA-LRM.

55. Receipts, January 31, 1815, and August 16, 1814, William Cramond receipt book 1814–1815, HSP. D'Arcy and Didier did business with David Curtis De Forest who had established the first American commercial house in Buenos Aires in 1815. De Forest provided political cover and served as agent for privateering activities. Jeffrey Orenstein, "Joseph Almeida: Portrait of a Privateer, Pirate & Plaintiff, Part II," *Green Bag 2d* 12 (2008): 36.

56. "Memorial on Behalf of the Owners of the Schooner Highflier, Having a Claim on the Mexican Government Addressed to the Board of Commissioners under the Convention of the 11th of April, 1839 between the United States of America and the Mexican Republic," Case Number 24, Thomas Tenant, Case Files for Cases Heard, 08/25/1850–1851, Record Group 76: Records of Boundary and Claims Commissions and Arbitrations, 1716–1994, National Archives, College Park, MD. Nixon Walker and George McCallmint to Roswell Lee, September 20, 1816, Box 1, Folder 8, SA-LRM.

57. William Cramond to Roswell Lee, July 15, 1815. Box 1, Folder 4, SA-LRM.

58. See, for example, William Cramond to Roswell Lee, July 17, September 2, September 25, September 28, 1815, April 11, 1817, Box 1, SA-LRM.

59. William Cramond to Roswell Lee, November 6, 1816, Box 1, Folder 9, SA-LRM.

60. William Cramond to Roswell Lee, June 12, 1817, Box 2, Folder 7, SA-LRM.

61. William Cramond to Roswell Lee, November 6, 1816, Box 1, Folder 9, Nixon Walker and George McCallmint to Roswell Lee, September 20, 1816, Box 1, Folder 8, and William Cramond, Philadelphia, to Roswell Lee, Superintendent, Springfield Armory, April 11, 1817, Box 2, Folder 6, SA-LRM. For quality issues, see Nixon Walker and George McCallmint to Roswell Lee,

Superintendent, Springfield Armory, September 20, 1816, Box 1, Folder 8, SA-LRM.

62. William G Miller to James Monroe, July 6, 1812, USDS-DC Buenos Aires. Gervasio Antonio de Posadas to James Madison, March 9, 1814, USDS-DC Buenos Aires. Posadas wrote to Madison: "The preponderance we should give *to your influence in the commercial world* has not a little influence. It is in you we place our present hopes."

63. Thomas Murray, *The Story of the Irish in Argentina* (New York: P. J. Kennedy and Sons: 1919); and Juan Martin de Pueyredón to James Monroe, January 31, 1817, USDS-DC Buenos Aires. Murray, *The Story of the Irish in Argentina*.

64. Thomas Lloyd Halsey to John Quincy Adams, November 31, 1817, USDS-DC Buenos Aires.

65. "List of vessels sailing from port of Valparaiso between July 1817 and July 1818" in USDS-DC Valparaiso.

66. For example, Hill alerted merchant correspondents that most vessels only paid a duty of 10 or 12 percent and that munitions and military stores were free, which influenced the decisions of exporters. Henry Hill to D'Arcy and Didier and Shepard, June 16, 1817, and Hill to Palmer and Hamilton, D'Arcy and Didier, January 30, 1818, Box 1, Folder 2, Henry Hill Papers, Yale University Manuscripts and Archives.

67. Henry Hill to D'Arcy and Didier and Shepard, February 6, 1817, June 22, 1817, and Hill to Palmer and Hamilton, D'Arcy and Didier, January 30, 1818, Box 1, Folder 2, Henry Hill Papers, Yale University Manuscripts and Archives.

68. Henry Hill to John Quincy Adams, September 28, 1819, USDS-DC Valparaiso.

69. Thomas Lloyd Halsey to John Quincy Adams, November 31, 1817, USDS-DC Buenos Aires. See also William G. Miller to John Quincy Adams, July 12, 1821, USDS-DC Montevideo. Miller similarly wrote about his favorable relations with generals and government officials.

70. The United States would, following President Monroe's message to Congress at the end of 1822, "sustain its neutral position, and allow to each party, while the war continues, equal rights . . . according to the well-known law of nations." US Senate Journal, "Annual Message to Congress," December 3, 1822, 17th Congress, 2nd Session: 16.

71. James Monroe, US Senate Journal, "Annual Message to Congress," December 2, 1823, 18th Congress, 1st Session: 11. For scholarly interpretations, see Dexter Perkins, *A History of the Monroe Doctrine* (Boston: Little, Brown, 1941); Weeks, *John Quincy Adams and American Global Empire*; and Jay Sexton, *The Monroe Doctrine: Empire and Nation in Nineteenth-Century America* (New York: Hill and Wang, 2011). While Monroe's message lacked

military teeth, it did informally bind the United States to take action should this declaration have been violated—an obligation to which Latin American officials held the US government accountable and to which it often complied. For examples of ministers asking the US State Department for clarification on its promise to resist foreign interference by the Holy Alliance and later requesting US military intervention against Britain and diplomatic intervention in negotiation with Spain, see José María Salazar to John Quincy Adams, July 2, 1824, USDS, Notes from the Colombian Legation, 1810–1906, RG59 National Archives, Washington, 1943. Microfilm, no-M51, roll 2. Domingo Acosta to Louis McLane, April 17, 1834, and Domingo Acosta to John Forsyth, February 25, 1837. See Pablo Obregon to Henry Clay (acknowledging the receipt of his note saying a US minister had been sent to Russia requesting its influence with Spain in concluding peace with Mexico), January 4, 1824, USDS, Notes from the Mexican Legation, 1821–1906, RG59 National Archives, Washington, 1960. Microfilm, no-M54, roll 1.

72. William G. D. Worthington to John Quincy Adams, January 15, 1818, USDS, Despatches from US Ministers in Argentina, RG59, National Archives, Washington, 1944. Microfilm, no. M-51, roll 1.

73. For ports in South America, see, for example, "A Statement showing the Amount of vessels into the Ports of the Republic of Venezuela for the year 1832" in despatch from J. G. A. Williamson to Louis Mclane, May 4, 1833, USDS-DC La Guiara. For the quotation, see John M. MacPherson to Martin Van Buren, December 14, 1829, USDS-DC Cartagena.

74. For the importance of the overland trade between the United States and Mexico, see Richard J. Salvucci, "The Origins and Progress of US-Mexican Trade, 1825–1884," *Hispanic American Historical Review* 71, no. 4 (Nov. 1991): 697–735. Salvucci argues that it increased immensely after 1825, pp. 701–702.

75. John H. Coatsworth and Jeffrey G. Williamson, "Always Protectionist? Latin American Tariffs from Independence to Great Depression," *Journal of Latin American Studies* 36, no. 2 (May 2004): 213–214; and Robert A. Potash, *Mexican Government and Industrial Development in the Early Republic: The Banco de Avio* (Amherst: University of Massachusetts Press, 1983).

76. Paul Gootenberg, "North-South: Trade Policy, Regionalism and Caudillismo in Post-Independence Peru," *Journal of Latin American Studies* 23, no. 2 (May 1991): 282–283.

77. D. C. M. Platt, "Dependency in Nineteenth-Century Latin America: An Historian Objects," *Latin American Research Review* 15, no. 1 (1980): 117.

78. See, for example, Joseph LeRoy Love and Nils Jacobsen, *Guiding the Invisible Hand: Economic Liberalism and the State in Latin American History* (Westport, CT: Praeger, 1988), and Stephen H. Haber, ed., *How Latin America Fell Behind: Essays on the Economic Histories of Brazil and Mexico, 1800–1914* (Stanford, CA: Stanford University Press, 1997).

79. Coatsworth and Williamson, "Always Protectionist?," 208.

80. William Taylor to John Quincy Adams, December 19, 1824, and January 5, 1825, USDS-DC Vera Cruz.

81. See Consular Return January 1–June 30, 1826, in William Taylor to Henry Clay, August 2, 1826, and letter January 8, 1826, USDS-DC Vera Cruz.

82. Platt, "Dependency in Nineteenth-Century Latin America," 117.

83. Paul Gootenberg, "The Social Origins of Protectionism and Free Trade in Nineteenth-Century Lima," *Journal of Latin American Studies* 14, no. 2 (Nov. 1982): 341–342.

84. Stockholders' Minutes 1822–1842, October 29, 1824, Merrimack Manufacturing Company Collection, BLHC.

85. New York markets were overstocked in the 1820s, and in bad economic years the Boston Associates relied on exports to South America. Caroline Ware, *The Early New England Cotton Manufacture, a Study in Industrial Beginnings* (Boston: Houghton Mifflin, 1931), 90, 190.

86. William Tudor to Nathan Appleton, May 3, 1824, Appleton Family Papers, MHS.

87. Northbridge Cotton Manufacturing Company Collection, Business Records of Various New England and New York Textile Firms, BLHC.

88. Thomas Lloyd Halsey to Israel Thorndike, August 21, 1818, Israel Thorndike Collection, BLHC.

89. John M. Forbes to John Quincy Adams, December 29, 1821, and January 21, 1822, USDS-DC Buenos Aires.

90. William Tudor to José de la Serna e Hinojosa, May 25, 1824, and August 30, 1824, USDS-DC Lima.

91. William Tudor to Nathan Appleton, May 3, 1824, Appleton Family Papers, MHS.

92. Robert F. Dalzell, *Enterprising Elite: The Boston Associates and the World They Made* (New York: W. W. Norton, 1987), 50. Accounts Current 1822–1831, Boston Manufacturing Co. Collection, BLHC. In August of 1824, for example, the company had in stock $164,623 of bleached and brown cloth, page 96. They would continue to manufacture large quantities of the cloth. See, for example, Accounts Current 1830–1840, February 1832, Boston Manufacturing Co. Collection, BLHC.

93. Entries, March–April 1825, Journal 1825, Vol. 29 Hamilton Manufacturing Co. Collection, BLHC. Accounts 1828–1831, Miscellaneous Papers, Hamilton Manufacturing Co. Collection, BLHC. They would specialize in flannels, drills, and dimities—lightweight, sheer cotton fabric. Dalzell, *Enterprising Elite,* 50.

94. William Tudor to Henry Clay, June 11, 1826, USDS-DC Lima.

95. William Tudor to José Manuel de Pardo, November 2, 1826, USDS-DC Lima.

96. William Tudor to Henry Clay, January 6, 1827, USDS-DC Lima.

97. In a letter to one of their selling agents, the Newmarket Manufacturing Company wrote that they had received information that cotton sheetings were selling well in Peru and Chile. Letter to Hacker Brown and Co., June 26, 1827, and October 15, 1828, Newmarket Manufacturing Company Vol. 1, Business Records of Various New England and New York Textile Firms, BLHC. The Peruvian production of tucuyos had been a problem for imports since at least the early 1820s, but by the time of Tudor's negotiations, American bleached and unbleached cottons had supplanted those of other countries. See Thomas Hockley, March 1822, Thomas Hockley Letterbook 1819–1822, Hockley family papers 1731–1883, HSP; and Charles Frederick Bradford, "South American Market 1826–1830," Edward Hickling Bradford Family Papers, MHS.

98. Barbara M. Tucker, *Samuel Slater and the Origins of the American Textile Industry, 1790–1860* (Ithaca: Cornell University Press, 1984), 91–109. Tucker argues that Slater could not adapt to the new market society, p.109.

99. Letter to Hacker Brown and Co., June 26, 1827, Newmarket Manufacturing Company Vol. 1, Business Records of Various New England and New York Textile Firms, BLHC.

100. John T. Mansfield to Henry Clay, July 15, 1826, USDS-DC Pernambuco.

101. Charles Frederick Bradford to Perkins and Company, July 10, 1826, Edward Hickling Bradford Family Papers, MHS.

102. Michael Hogan to Edward Livingston, May 5, 1832, USDS-DC Valparaiso.

103. Letter to Kimball and Clark, January 10, 1828, Newmarket Manufacturing Company Vol. 1, Business Records of Various New England and New York Textile Firms, BLHC.

104. See, for example, "Report of the Vessels of the United States which Enter at the Port of Guayaquil from July 1 to December 31, 1825," USDS-DC Guayaquil; "Report of the Arrival, Departure, Cargo, Outward and Inward, of American Vessels Arriving at the Port of La Guiara, July 1 to December 31, 1826," USDS-DC La Guiara; and "Consular Return January to June 1826" in William Taylor to Henry Clay, August 2, 1826, USDS-DC Vera Cruz.

105. "Minutes of a Conference with Mr. Rivadavia, Buenos Aires," September 17, 1821, USDS-DC Buenos Aires.

106. David Curtis De Forest to Brown and Ives, January 15, 1821, Box 4, Folder 18, Letterbook 7, De Forest Family Papers, Yale Collection of American Literature, Beinecke Rare Book and Manuscript Library.

107. J. G. A. Williamson to P. Dias, Secretary of Exterior Relations Caracas, February 3, 1834, and March 3, 1834, USDS-DC La Guiara.

108. For information on the claims commission that commenced in 1839 to settle American claims against the Mexican government, see the digital work of

David Mackenzie, George Mason University, http://davidmckenzie.info/claims/index.php.

109. See, for example, William Davis Robinson, *Memoirs of the Mexican Revolution: including a narrative of the expedition of General Xavier Mina* (London: Lackington, Hughes, Mavord & Lepard, 1821), 62, 252, 365; and William Taylor to Henry Clay, January 18, 1826, USDS-DC Vera Cruz.

110. For a rise in the overland trade between the United States and Mexico after 1825, see Salvucci, "The Origins and Progress of US-Mexican Trade," 701–702.

111. William Taylor to John Quincy Adams, June 4, 1824, USDS-DC Vera Cruz.

112. William Taylor to Henry Clay, November 18, 1825, and John Wilkinson to William Taylor, October 25, 1825, USDS-DC Vera Cruz.

113. Historian D. K. Fieldhouse has characterized the exploitation of natural resources as the defining characteristic of colonial empires. Fieldhouse, *The Colonial Empires: A Comparative Survey from the Eighteenth Century*, 2nd ed. (London: Palgrave Macmillan, 1982), 386.

114. John Ward to Martin Van Buren, November 30, 1830, USDS-DC Chihuahua. Similarly, customs officials in Panama City, according to the US consul, illegally detained American cargoes of specie, perhaps because of this resentment over exportation. J. B. Ferand to Governor of Panama, June 14, 1834, USDS-DC Panama City.

115. For the extensive exportation of Mexican logwood to the United States, see, for example, the multiple consular returns contained in USDS-DC Campeche. See also Salvucci, "The Origins and Progress of US-Mexican Trade," 701–702. For examples of manufacturing companies that relied on imports of logwood see Return for Athol Manufacturing Company, Schedules for Maine and New Hampshire, US Department of State, *Records of the 1820 Census of Manufactures*, RG279 National Archives, Washington, 1944, roll 2; Slater and Tiffany, Order Book, August 9, 1824, Slater Companies Collection, BLHC; Merrimack Manufacturing Company, Journal 1826–1829, page 239, Merrimack Manufacturing Company Collection, BLHC; and Aza Arnold, Inventory of dyestuffs, February 13, 1837, Folder 2 Memoranda, Aza Arnold Papers, Ms 266, RIHS.

116. See, for example, "Account of the Shipping Cargoes Outward and Inward of American Vessels," in J. G. A. Williamson to Henry Clay, January 1, 1828, and "Statement of American Vessels, January 1–July 1, 1830," in J. G. A. Williamson to Martin Van Buren, July 8, 1830, USDS-DC La Guiara. Also, "Consular Return July 1–December 30, 1835," in John Patrick to John Forsyth, December 31,1835, and "Consular Return January 1–June 30, 1836," in John Patrick to John Forsyth, July 20, 1836, USDS-DC Montevideo.

117. "Mexico and Mr. Poinsett: Reply to a British Pamphlet, entitled

'Observation on the Instructions given by the president of the United States of America to the representatives of that republic, at the Congress of Panama' " (Philadelphia, 1829), 11.

118. Kirshner, "Political Economy in Security Studies," 72.

119. Sheinin, "The New Dollar Diplomacy," 82.

120. The Colt Manufacturing Company, for example, was founded in Hartford in 1836 and would go on to be a major government contractor and exporter of firearms. And the Boston Associates and the Slater family established new factories throughout the 1830s.

Chapter 6. Industrial Manifest Destiny

1. Richard Rush to Joel Roberts Poinsett, January 13, 1838, Folder 17, Box 9, JRPP.

2. "Annexation," *The United States Magazine, and Democratic Review* (Washington: Langtree and O'Sullivan; J. & H. G. Langley, July/August 1845).

3. For "Manifest Destiny" as an American ideology from settlement to the twentieth century, see Anders Stephanson, *Manifest Destiny: American Expansionism and the Empire of Right* (New York: Hill and Wang, 1995). Robert W. Tucker and David C. Hendrickson argue that Thomas Jefferson's decision to purchase Louisiana from Napoleon reflected his belief in a nation that would expand throughout the continent to create new republics that would be governed by liberty and republican virtues. Tucker and Hendrickson, *Empire of Liberty: The Statecraft of Thomas Jefferson* (New York: Oxford University Press, 1990). For expansion as a racial mission, see Reginald Horsman, *Race and Manifest Destiny: The Origins of American Racial Anglo-Saxonism* (Cambridge: Harvard University Press, 1981). Amy S. Greenberg argues that by 1848 westward expansion was driven by two competing masculinities—"martial" versus "restrained" manhood. Greenberg, *Manifest Manhood and the Antebellum American Empire* (New York: Cambridge University Press, 2005), 9. Robert Pierce Forbes, *The Missouri Compromise and its Aftermath: Slavery & the Meaning of America* (Chapel Hill: University of North Carolina Press, 2007); Leonard Richards, *The Slave Power: The Free North and Southern Domination, 1780–1860* (Baton Rouge: Louisiana State University Press, 2000); and Kinley J. Brauer, *Cotton versus Conscience: Massachusetts Whig Politics and Southwestern Expansion, 1843–1848* (Lexington: University of Kentucky Press, 1967).

4. As Patricia Limerick reminds us, the settlement of the North American continent was about more than individual adventurism and violent bravery; it had everything to do with the brutal realities of capitalism. Patricia Nelson Limerick, *The Legacy of Conquest: The Unbroken Past of the American West* (New York: W. W. Norton, 1987).

5. Two of the most compelling are perhaps William Cronon's *Nature's Metropolis: Chicago and the Great West* (New York: W. W. Norton, 1992) and

Richard White's *Railroaded: The Transcontinentals and the Making of Modern America* (New York: W. W. Norton, 2011).

6. A board of military officers declared the arms produced by the federal government superior to solicitations from private firms. ASP, "Report of the President of a Board of Officers on Improvements in Fire-Arms by Hall, Colt, Cochran, Leavitt, and Baron Hackett, as Compared with the United States Musket," Military Affairs 7, no. 743 (Oct. 3, 1837): 525.

7. Cotton manufactures, however, increased almost threefold over the decade. United States Secretary of the Interior, *Manufactures of the United States in 1860* (Washington: GPO, 1865), xv.

8. Matthew Karp, however, acknowledges the ironic pairing of pro-slavery, states' rights policies with aggressive national militarism. Matthew J. Karp, "Slavery and American Sea Power: The Navalist Impulse in the Antebellum South," *Journal of Southern History* 77, no. 2 (May 2011): 283–324. John M. Belohlavek describes a John Tyler "triumvirate" that included two northerners with changeable views on expansion: Daniel Webster and Caleb Cushing. John M. Belohlavek, *Broken Glass: Caleb Cushing & the Shattering of the Union* (Kent, OH: Kent State University Press, 2005), 143.

9. Edward P. Crapol, "John Tyler and the Pursuit of National Destiny," *Journal of the Early Republic* 17, no. 3 (Autumn 1997): 469.

10. For the short-lived Aroostook War, see Howard Jones, "Anglophobia and the Aroostook War," *New England Quarterly* 48, no. 4 (Dec. 1975): 519–539.

11. ASP, "Report from the Secretary of the Treasury on the State of Finances," 29th Congress, 1st Session, no. 2 (Dec. 3, 1845): 14–15; and ASP, "Report of the Secretary of the Treasury on the State of Finances," 31st Congress, 2nd Session, no. 4 (Dec. 16, 1850): 117–118.

12. John M. Belohlavek, "'Let the Eagle Soar!': Democratic Constraints on the Foreign Policy of Andrew Jackson," *Presidential Studies Quarterly* 10, no. 1 (Winter 1980): 41; and ASP, Journal of the Senate, 22nd Congress, 2nd Session, no. 22 (Dec. 4, 1832): 8.

13. James Ellsworth De Kay, *Sketches of Turkey in 1831 and 1832* (New York: Harper, 1833), 191–192, 495.

14. About 90 percent of manufactured cotton was consumed domestically in 1840. *Executive Documents Printed by Order of the House of Representatives, During the Third Session of the Thirty-Fourth Congress, 1856–1857* (Washington: Cornelius Wendell, 1857), 210.

15. Even Texas and Mexico, combined, lagged far behind. ASP, "Report from the Secretary of the Treasury on the State of Finances," 29th Congress, 1st Session, no. 2 (Dec. 3, 1845): 14–15.

16. For the overseas implications of Manifest Destiny, see Brian, "Maritime Destiny as Manifest Destiny: American Commercial Expansionism and the Idea of the Indian," *Journal of the Early Republic* 30, no. 3 (Fall 2010): 377–411.

17. For the importance of the Pacific Ocean for American history, see David Igler, *The Great Ocean: Pacific Worlds from Captain Cook to the Gold Rush* (New York: Oxford University Press, 2013).

18. James P. Ronda, *Astoria and Empire* (Lincoln: University of Nebraska Press, 1990), 30–31.

19. James R. Fichter, *So Great a Profit: How the East Indies Trade Transformed Anglo-American Capitalism* (Cambridge: Harvard University Press, 2010), 226–230. Americans also traded opium: Jacques M. Downs, "American Merchants and the China Opium Trade, 1800–1840," *Business History Review* 42, no. 4 (Winter 1968): 418–442.

20. Frederick Merk, "The Oregon Question in the Webster-Ashburton Negotiations," *Mississippi Valley Historical Review* 43, no. 3 (Dec. 1956): 383.

21. ASP, "Message from the President of the United States to the Two Houses of Congress," 28th Congress, 1st Session (Dec. 5, 1843): 3.

22. ASP, "Petition of a Number of Citizens of the Territory of Oregon," 28th Congress, 1st Session, no. 105 (Feb. 7, 1844).

23. John H. Schroeder, "Rep. John Floyd, 1817–1829: Harbinger of Oregon Territory," *Oregon Historical Quarterly* 70, no. 4 (Dec. 1969): 333–346.

24. ASP, "Memorial of a Number of Citizens of Michigan Praying a Donation of land to emigrants and settlers in the Oregon Territory," 25th Congress, 3rd Session, no. 266 (Feb. 26, 1839).

25. Israel D. Andrews, *Communication from the Secretary of the Treasury* (Washington: Robert Armstrong, 1853), 236.

26. ASP, "Settle Oregon," 27th Congress, 3rd Session, no. 157 (Feb. 9, 1843): 8.

27. Fichter, *So Great a Profit*, 210–11, 285.

28. Kenneth Pomeranz, *The Great Divergence: China, Europe, and the Making of the Modern World Economy* (Princeton: Princeton University Press, 2000), 141.

29. Ernest R. May and John K. Fairbank, eds., *America's China Trade in Historical Perspective: The Chinese and American Performance* (Cambridge: Harvard University Press, 1986), 109–110.

30. *Report from the Select Committee of the House of Lords of the Affairs of the East India Company and into the Trade Between Great Britain, The East Indies and China* (London: House of Commons, 1830), 355, 366; and ASP, "China Trade," 26th Congress, 1st Session, no. 248 (July 1, 1840): 11.

31. Caroline F. Ware, *The Early New England Cotton Manufacture, a Study in Industrial Beginnings* (Boston: Houghton Mifflin, 1931), 194.

32. "Particulars of Goods shipped to Sundry Places on Account of Hamilton Manufacturing Co.," Accounts, 1829–1831, Volume 19, Hamilton Manufacturing Co. Collection, BLHC.

33. United States Secretary of the Interior, *Manufactures of the United States in 1860* (Washington: GPO, 1865), xix.

34. "Agreement between the owners, master, seamen and mariners of the Ship Mentor," 1819, Folder 3, Box 1, George Howe Papers, MHS.

35. Edwin Conant to George Howe, March 15, 1834, Folder 4, Box 1, and William Whitmore to George Howe, May 15, 1835, Folder 4, Box 1, George Howe Papers, MHS.

36. "Hooksett Manufacturing Company Report for the Week Ending February 9, 1839," Folder 8, Box 1, and "Cost Cotton delivered at Dover from May 1840 to May 1847," Folder 9, Box 1, George Howe Papers, MHS.

37. For debates among leading textile manufacturers about tariffs and the expansion of slavery, see Robert F. Dalzell, *Enterprising Elite: The Boston Associates and the World They Made* (New York: W. W. Norton, 1987), 194–204. Two classic studies of New England textile manufacturers and merchants who compromised morality for the sake of economic interests and unionism in their political support of the institution of slavery are Kinley J. Brauer, *Cotton versus Conscience: Massachusetts Whig Politics and Southwestern Expansion, 1843–1848* (Lexington: University of Kentucky Press, 1967), and Thomas H. O'Connor, *Lords of the Loom: The Cotton Whigs and the Coming of the Civil War* (New York: Scribner, 1968). For the connection between industrialization and plantation slavery, see Sven Beckert and Seth Rockman, eds., *Slavery's Capitalism: A New History of American Economic Development* (Philadelphia, University of Pennsylvania Press, 2016); Sven Beckert, *Empire of Cotton: A Global History* (New York: Vintage, 2015); and Walter Johnson, *River of Dark Dreams: Slavery and Empire in the Cotton Kingdom* (Cambridge: Harvard University Press, 2013).

38. This opium trade was limited to a small number of firms, and it faced difficulties from Chinese attempts to curb the trade. It was only about a tenth the size of Britain's opium trade, which made up about two-thirds of all its imports into China. Jacques M. Downs, "American Merchants and the China Opium Trade, 1800–1840," *Business History Review* 42, no. 4 (Winter 1968): 418–442.

39. Belohlavek, *Broken Glass*, 153.

40. For Cushing's role, see Belohlavek, *Broken Glass*, 150–180.

41. Belohlavek, *Broken Glass*, 183.

42. Belohlavek, *Broken Glass*, 162, 183.

43. Dalzell, *Enterprising Elite*, 61; Perkins shipped furs to Smyrna to purchase opium to sell in China. James and T. H. Perkins Account book, September 26, 1823, page 39, Reel 7, Vol. 40, 1822–1823, Thomas Handasyd Perkins Papers, MHS.

44. ASP, "Report of the Secretary of the Treasury on the State of Finances," 31st Congress, 2nd Session, no. 4 (Dec. 16, 1850): 117.

45. ASP, "Senate Committee Report on the Post Office and Post Roads," 29th Congress, 1st Session, no. 306 (March 5, 1846): 3–5. According to another

report, the fur trade was "of minor importance when compared with the great object of encouraging, facilitating, and guarding emigration to the Pacific." ASP, "Military Posts on the Route to Oregon," 29th Congress, 1st Session, no. 13 (Dec. 31, 1845): 3.

46. ASP, "Settle Oregon," 27th Congress, 3rd Session, no. 157 (Feb. 9, 1843): 7.

47. ASP, "Military Posts on the Route to Oregon," 29th Congress, 1st Session, no. 13 (Dec. 31, 1845): 2.

48. This phrase is often linked to the election of 1844, but it did not appear until 1846. Hans Sperber, "'Fifty-Four Forty or Fight': Facts and Fictions," *American Speech* 32, no. 1 (Feb. 1957): 5–11.

49. Amy S. Greenberg, *A Wicked War: Polk, Clay, Lincoln and the 1846 US Invasion of Mexico* (New York: Knopf, 2012), 97.

50. ASP, "Message from the President, Transmitting Copies of Correspondence between the State Department and Late Commissioner to China," 33rd Congress, 1st Session, no. 123 (July 19, 1854): 208.

51. ASP, "Railroad to Oregon," 30th Congress, 1st Session, no. 733 (June 23, 1848): 9.

52. ASP, "Report from the Secretary of the Treasury on the State of Finances," 29th Congress, 1st Session, no. 2 (Dec. 3, 1845): 403; and Dalzell, *Enterprising Elite,* 52. Tyler actually supported the 1842 tariffs. Crapol, "John Tyler," 480.

53. See, for example, "Resolutions of the General Assembly of Connecticut, Relative to the Tariff," 28th Congress, 2nd Session, no. 3 (Dec. 10, 1844).

54. Boston Board of Trade, *Third Annual Report of the Government* (Boston: George C. Rand and Avery, 1857), 172–173.

55. The tariff debate largely broke down along sectional lines, but there were exceptions, with Virginia favoring protection and New Hampshire favoring a reduction in rates. Some of the Associates had supported the annexation of Texas hoping to get Southern planters to "compromise" on protectionism.

56. Congressional Globe, House of Representatives, 29th Congress, 1st Session (June 18, 1846), 990.

57. ASP, "Report of the Secretary of the Treasury on the State of Finances," 31st Congress, 2nd Session, no. 4 (Dec. 16, 1850): 18; and Douglas A. Irwin and Peter Temin, "The Antebellum Tariff on Cotton Textiles Revisited," *Journal of Economic History* 61, no. 3 (Sept. 2001): 777–798.

58. US Secretary of the Interior, *Manufactures of the United States in 1860* (Washington: GPO, 1865), xv.

59. This was just as he had four years earlier, when he learned that Mexico planned to invade Texas. Joel Roberts Poinsett to John C. Hamilton, May 26, 1842, Folder 13, Box 16, JRPP.

60. Joel Roberts Poinsett to Lewis Cass, May 16, 1846, Folder 17, Box 16, JRPP.

61. Gouverneur Kemble to Joel Roberts Poinsett, October 21, 1847, Folder 18, Box 16, JRPP.

62. ASP, "Report from the Secretary of War," 25th Congress, 3rd Session, no. 273 (Feb. 28, 1839): 2, 554.

63. Richard Rush to Joel Roberts Poinsett, February 18, 1838, Folder 5, Box 10, JRPP.

64. James Renelden to Joel Roberts Poinsett, March 2, 1838, Folder 7, Box 10, JRPP.

65. ASP, "Report of the President of a Board of Officers on Improvements in Fire-Arms by Hall, Colt, Cochran, Leavitt, and Baron Hackett, as Compared with the United States Musket," Military Affairs 7, no. 743 (Oct. 3, 1837): 528.

66. ASP, "Documents Accompanying the Report of the Secretary of War," no. 1 (Nov. 21, 1840): 58.

67. Hamilton Wilcox Merrill, February 26, 1844, Private Book, Vol. 2, HM4022, Merrill Collection, Huntington Library, San Marino, CA.

68. Nathan Starr to John Rodgers, President of the Board of Navy Commissioners, March 23, 1816, Volume 3, Records Collection of the Office of Naval Records and Library, Record Group 25, Entry 328, National Archives Building, Washington.

69. Asa H. Waters, *Biographical Sketch of Thomas Blanchard and His Inventions* (Worcester, MA: L. P. Goddard, 1878), 3–4.

70. Dwight Goddard, *A Short Story of Thomas Blanchard* (Worcester, MA: Wyman-Gordon, 1900), 4.

71. Carolyn C. Cooper, *Shaping Invention: Thomas Blanchard's Machinery and Patent Management in Nineteenth-Century America* (New York: Columbia University Press, 1991), 17–18.

72. Waters, *Biographical Sketch*, 6.

73. Merritt Roe Smith, *Harpers Ferry Armory and the New Technology: The Challenge of Change* (Ithaca: Cornell University Press, 1977), 127.

74. Charles R. Morris, *The Tycoons: How Andrew Carnegie, John D. Rockefeller, Jay Gould, and J. P. Morgan Invented the American Supereconomy* (New York: Times Books, 2005), 34–35.

75. Smith, *Harpers Ferry Armory and the New Technology*, 134–135.

76. Morris, *The Tycoons*, 37.

77. Cooper, *Shaping Invention*, 116.

78. Steven Lubar, "The Transformation of Antebellum Patent Law," *Technology and Culture* 32 (Oct. 1991), 957–958.

79. Jonathan Amory to Francis C. Lowell, July 10, 1836, Folder 5.6, Box 6, Francis Cabot Lowell II Papers, MHS.

80. Jessica Lepler, *The Many Panics of 1837: People, Politics, and the Creation of a Transatlantic Financial Crisis* (New York: Cambridge University Press, 2013).

81. Lubar, "The Transformation."

82. The board reported that "however ingenious therefore may be the invention however credible the skill of the manufacturer, the board is of the opinion that the arm of Cochran is an unsafe weapon . . . the arm of Colt may be usefully applied in special cases . . . naval service . . . but not adapted to the general purposes of the service." ASP, "Report of the President of a Board of Officers on Improvements in Fire-Arms by Hall, Colt, Cochran, Leavitt, and Baron Hackett, as Compared with the United States Musket," Military Affairs 7, no. 743 (Oct. 3, 1837): 528.

83. Gouverneur Kemble to Joel Roberts Poinsett, September 10, 1840, Folder 7, Box 15, JRPP.

84. ASP, "Report of the President of a Board of Officers on Improvements in Fire-Arms," 528.

85. ASP, "Report of the President of a Board of Officers on Improvements in Fire-Arms," 529.

86. ASP, "Documents from War Department," 24th Congress, 2nd Session, no. 2 (Nov. 1, 1836): 328; M. W. Edwards to Asa Waters, December 15, 1834, Box W 3, Waters Family, Papers, 1749–1873, AAS.

87. Contract with Lemuel Pomeroy, February 24, 1840, Volume 2, Contracts for Ordnance and Ordnance Supplies, Records of the Chief of the Ordnance Department, Record Group 156, Entry 78, National Archives, Washington.

88. Asa Waters to George Talcott, November 14, 1840; Asa Waters to George Bomford, August 28, 1841, Folio Volume 1, Waters Family, Papers, AAS.

89. Asa Waters to Eli Whitney Jr., December 8, 1845, Octavo Volume 7, Letterbook 1837–1865, Waters Family, Papers, AAS.

90. Thomas Tyson, "Accounting for Labor in the Early Nineteenth Century: The US Arms Making Experience," *Accounting Historians Journal* 17, no. 1 (June 1990): 54.

91. ASP, "On Claim of John H. Hall for Compensation," 25th Congress, 2nd Session, no. 756 (Jan. 4, 1838).

92. Pamela Haag, *The Gunning of America: Business and the Making of American Gun Culture* (New York: Basic, 2016), 28.

93. Henry Barnard, *Armsmear: The Home, the Arm, and the Armory of Samuel Colt: A Memorial* (New York: Alvord, 1860), 166–168. For soldiers' difficulties reloading their guns in Florida, see "Scraps Preserved by N. M. for H. W. M.," April 3, 1840, HM4020, Merrill Collection, Huntington Library, San Marino, CA, and Jacob Neff, *The Army and Navy of America: From the Period of the French and Indian Wars to the Close of the Florida War* (Philadelphia, 1845), 610.

94. Secretary of War Joel Poinsett, for example, said he didn't want to waste any more government resources on trying to increase the rapidity of firing

"without long-tried experiments in the field," ASP, "Report of the Secretary of War" (Dec. 5, 1840): 21.

95. ASP, "Report from the Secretary of War, transmitting the report of a board of dragoon officers appointed to witness an exhibition of the repeating fire-arms and water-proof ammunition invented by Samuel Colt," 26th Congress, 2nd Session, no. 14 (Dec. 16, 1840): 2–3.

96. Herbert C. Houze, "The Paterson Era," in Elizabeth Mankin Kornhauser, ed., *Samuel Colt: Arms, Art, and Invention* (New Haven: Yale University Press, 2006), 56.

97. Pamela Haag argues that civilian consumption of patent firearms, such as Colt's, was limited before the Civil War. Haag, *The Gunning of America*, 58–59.

98. US Circuit Court, *The trial of Samuel Colt: complete report of the trial of Samuel Colt vs. the Mass. Arms Company, tried June 30, 1851, in U.S. Circuit Court, Boston, Mass.* (Harriman, TN, 1853), 8. Colt's assertion reflected his own experience, but was slightly hyperbolic. For example, Eliphalet Remington of Suffield, CT, founded E. Remington and Sons in New York in 1816 and had some success selling flintlock rifles and other firearms to civilian consumers in the 1820s and 1830s. Roy Marcot, *History of Remington Firearms: The History of One of The World's Most Famous Gun Makers* (Guilford, CT: Globe Pequot Press, 2005).

99. Tyson, "Accounting for Labor," 49.

100. Charles Stearns, *The National Armories: A Review of the Systems of Superintendency, Civil and Military, Particularly with Reference to Economy and General Management at the Springfield Armory* (Springfield, MA: George Wilson's Steam Power Presses, 1853), 13, 74.

101. M. L. Brown, "Notes on US Arsenals, Depots, and Martial Firearms of the Second Seminole War," *Florida Historical Quarterly* 61, no. 4 (April 1983): 453.

102. ASP, "Documents from War Department," 27th Congress, 3rd Session, no. 2 (Nov. 14, 1842): 208.

103. Haag, *The Gunning of America*, 28.

104. ASP, "Report of the Secretary of War," 29th Congress, 1st Session, no. 344 (May 13, 1846): 2.

105. Brown, "Notes on US Arsenals," 449–451; ASP, "Expenditures in Suppressing Indian Hostilities in Florida," 26th Congress, 2nd Session, no. 8 (Dec. 15, 1840): 9.

106. Adams as president attempted to purchase Texas from Mexico for $1 million (1827) and Jackson imitated the action, but he offered $5 million in 1829. Both efforts were rebuffed by Mexico.

107. Crapol, "John Tyler," 469.

108. Paul E. Sturdevant, "Robert John Walker and Texas Annexation: A Lost Champion," *Southwestern Historical Quarterly* 109, no. 2 (Oct. 2005): 196. Britain had long been developing an informal empire in Mexico that threatened US interests there. Robert D. Aguirre, *Informal Empire: Mexico and Central America in Victorian Culture* (Minneapolis: University of Minnesota Press, 2004).

109. R. A. McLemore, "The Influence of French Diplomatic Policy on the Annexation of Texas," *Southwestern Historical Quarterly* 43, no. 3 (Jan. 1940): 342–347.

110. Ward McAfee, "A Reconsideration of the Origins of the Mexican-American War," *Southern California Quarterly* 62, no. 1 (Spring 1980): 49–65.

111. Greenberg, *Manifest Destiny,* 103–108.

112. [Philadelphia] *North American*, May 23, 1846, p.1.

113. ASP, "List of Papers Accompanying the Report of the Secretary of War," no. 1 (Dec. 5, 1846): 165.

114. ASP, "List of Papers Accompanying the Report of the Secretary of War," no. 8 (Nov. 30, 1847): 686.

115. Brown, "Notes on US Arsenals," 449–451; ASP, "Expenditures in Suppressing Indian Hostilities in Florida," 26th Congress, 2nd Session, no. 8 (Dec. 15, 1840): 9.

116. ASP, "Presidential message communicating information on mode of raising funds for carrying on war with Mexico," 29th Congress, 1st Session, no. 392 (June 16, 1846): 11.

117. ASP, "Message from the President of the United States," 29th Congress, 1st Session, no. 392 (June 16, 1846): 3–4.

118. Memorial of Private Contractors to US Congress [1835?], Waters Family, Papers, 1749–1873, Box W1, Folder 4 1835, AAS.

119. Contract with Simeon North, May 2, 1839, Volume 2, Records of the Chief of the Ordnance Department, Record Group 156, Entry 78, National Archives, Washington.

120. ASP, "List of Papers Accompanying the Report of the Secretary of War," no. 162 (Dec. 5, 1846).

121. Robert B. Gordon, "Simeon North, John Hall, and Mechanized Manufacturing," *Technology and Culture* 30, no. 1 (Jan. 1989): 179–188, and "Who Turned the Mechanical Ideal into Mechanical Reality?" *Technology and Culture* 29, no. 4 (Oct. 1988): 744–778.

122. Merritt Roe Smith, "John H. Hall, Simeon North, and the Milling Machine: The Nature of Innovation among Antebellum Arms Makers," *Technology and Culture* 14, no. 4 (Oct. 1973): 588.

123. Paul J. Uselding, "Henry Burden and the Question of Anglo-American Technological Transfer in the Nineteenth Century," *Journal of Economic History* 30, no. 2 (June 1970): 312–337.

124. Smith, "John H. Hall, Simeon North, and the Milling Machine," 589.

125. Asa Waters to Eli Whitney Jr., December 8, 1845, Octavo Volume 7, Letterbook 1837–1865, Waters Family, Papers, AAS.

126. ASP, "Petition of Samuel Colt," 30th Congress, 2nd Session, no. 3 (Dec. 12, 1848): 1–3.

127. David E. Twiggs to Thomas Jefferson Rusk, April 21, 1848, CHS, *Samuel Colt's Own Record of Transactions with Captain Walker and Eli Whitney, Jr., in 1847* (Hartford, CT, 1949), 84–85.

128. James B. Whisker, *The Rise and Decline of the American Militia System* (Selinsgrove, PA: Susquehanna University Press, 1999), 333.

129. ASP, "List of Papers Accompanying the Report of the Secretary of War," no. 8 (Nov. 30, 1847): 133–134, 166–167, 190.

130. [Philadelphia] *The North American,* May 19, 1846, p.1.

131. John C. Calhoun, Speech to Congress on Mexico, January 4, 1848, in Ernesto Chávez, ed., *The US War with Mexico: A Brief History with Documents* (Boston: Bedford/St. Martin's, 2008), 118–120.

132. For treaty negotiations, see Greenberg, *A Wicked War*, 256–260.

133. Joel H. Silbey, *Storm over Texas: The Annexation Controversy and the Road to Civil War* (New York: Oxford University Press, 2007), 96.

134. ASP, "Approval of the Oregon Bill," 30th Congress, 2nd Session, no. 3 (Dec. 6, 1848).

135. ASP, "Memorial of J. Quinn Thornton, praying the establishment of a Territorial Government in Oregon," 30th Congress, 1st Session, no. 143 (May 25, 1848): 15, 23.

136. Asa Waters to Eli Whitney Jr., July 10, 1846, Octavo Volume 7, Letterbook 1837–1865, Waters Family, Papers, AAS.

137. ASP, "Message from the President, Transmitting Copies of Correspondence between the State Department and Late Commissioner to China," 33rd Congress, 1st Session, no. 123 (July 19, 1854): 108.

138. ASP, "Report on the commercial relations of the United States with all foreign nations," 34th Congress, 1st Session, no. 47 (Dec. 3, 1855): 726–727.

139. *Executive Documents Printed by Order of the House of Representatives, During the Third Session of the Thirty-Fourth Congress, 1856–1857* (Washington: Cornelius Wendell, 1857), 210.

140. [Alexandria, VA] *Alexandria Gazette*, June 17, 1846, p. 3.

141. "Report of Mr. Poinsett on Tendering his Resignation," March 2, 1841, Folder 10, Box 16, JRPP.

142. ASP, "Report of the Secretary of the Treasury, showing the receipts and expenditures, &c., for the fiscal year ending June 30, 1850," 31st Congress, 2nd Session, no. 11 (Dec. 16, 1850): 117. This trade built on decades of increased American commerce on the Chihuahua and Santa Fe Trails. Max L. Moorhead,

New Mexico's Royal Road: Trade and Travel on the Chihuahua Trail (Norman: University of Oklahoma Press, 1954).

143. ASP, "Report of the Secretary of the Treasury, showing the receipts and expenditures, &c., for the fiscal year ending June 30, 1850," 31st Congress, 2nd Session, no. 11 (Dec. 16, 1850): 65. The federal government's ability to protect settlers had long mattered for its relationship with frontier settlers. American military and commercial benefits encouraged residents of New Mexico to acquiesce to the United States during the Mexican-American War. Moorhead, *New Mexico's Royal Road,* 193.

144. ASP, "List of Documents Accompanying the Report of the Secretary of War," no. 2 (Nov. 21, 1851, and June 23, 1851): 161, 328–329.

145. ASP, "List of Documents Accompanying the Report of the Secretary of War," no. 2 (Oct. 28, 1851): 448.

146. Haag, *The Gunning of America*. Colt excelled at branding and successfully employed New York wholesalers to distribute his arms nationwide. Harold C. Livesay, "Marketing Patterns in the Antebellum American Iron Industry," *Business History Review* 45, no. 3 (Autumn 1971): 286.

147. ASP, "Petition of Samuel Colt," 30th Congress, 2nd Session, no. 3 (Dec. 12, 1848): 2.

148. [Hartford, CT] *Courant*, January 20, 1849, p. 10.

149. [Salem, MA] *Salem Register*, October 4, 1847, p. 2.

150. Richard A. Dillio, "Samuel Colt's Peacemaker: The Advertising that Scared the West," *History of Media Technology* (Dec. 2017): 23.

Conclusion

1. Doron Ben-Atar, *Trade Secrets: Intellectual Piracy and the Origins of American Industrial Power* (New Haven: Yale University Press, 2004), 210–221.

2. US Secretary of the Interior, *Manufactures of the United States in 1860* (Washington: GPO, 1865), xxxiii.

3. Andrew Elmer Ford, *History of the Origin of the Town of Clinton, Massachusetts, 1653–1865* (Clinton, MA: W. J. Coulter, 1896), 231–232.

4. Nathan Rosenberg, *Technology and American Economic Growth* (New York: Harper & Row, 1972), 87–88. David Hounshell notes that while Rosenberg and others attribute the expression to a variety of British reports on American manufacturing in the mid-1850s, it was not really used except by historians, and not until the early twentieth century. David Hounshell, *From the American System to Mass Production, 1800–1932: The Development of Manufacturing Technology in the United States*, vol. 4 (Baltimore: Johns Hopkins University Press, 1985), 16–17.

5. Hounshell, *From the American System to Mass Production*, 16–25.

6. ASP, "Furniture for the President's House," 27th Congress, 2nd Session, no. 552 (April 1, 1842).

7. Mark R. Wilson, *The Business of Civil War: Military Mobilization and the State, 1861–1865* (Baltimore: Johns Hopkins University Press, 2006), 2, 35.

8. Paul A. C. Koistinen, *Beating Plowshares into Swords: The Political Economy of American Warfare, 1606–1865* (Lawrence: University Press of Kansas, 1996), 76.

9. Arms produced by the federal government were declared superior to solicitations from private firms. ASP, "Report of the President of a Board of Officers on Improvements in Fire-Arms by Hall, Colt, Cochran, Leavitt, and Baron Hackett, as Compared with the United States Musket," Military Affairs 7, no. 743 (Oct. 3, 1837): 525.

10. For frustration with the cessation of contract payments, see Asa H. Waters to Eli Whitney Jr., December 8, 1845, Octavo Volume 7, Letterbook 1837–1865, Waters Family, Papers, AAS.

11. Waters sold arms to Mexico in the 1840s. Asa H. Waters and Co. to Richard M. Jones, October 13, 1842, Octavo Volume 7, Letterbook 1837–1865, Waters Family, Papers, AAS.

12. See, for example, Samuel Colt Contract for Furnishing Rifles, February 4, 1850, and E. Remington and Sons Contract for Rifles, November 21, 1851, Contracts for Ordnance and Ordnance Supplies, Volume 3, Records of the Chief of the Ordnance Department, Record Group 156, Entry 78, National Archives, Washington.

13. Koistinen, *Beating Plowshares into Swords,* 140–141.

14. Chile's total imports from the United States increased from $1,244,171 in 1848 to $1,911,479 in 1850 to $4,594,911 in 1851. ASP, "Report on the commercial relations of the United States with all foreign nations," 34th Congress, 1st Session, no. 47 (Dec. 3, 1855): 726–734.

15. Amos Binney to Daniel Parker, March 13, 1845, Folder 5, Box 12, Daniel Parker Papers (Collection 466), HSP.

16. Alexander Hamilton, *Report on Manufactures*, December 5, 1791, in Harold C. Syrett, ed., *The Papers of Alexander Hamilton,* vol. 19 (New York: Columbia University Press, 1973).

17. Eliga H. Gould, *Among the Powers of the Earth: The American Revolution and the Making of a New World Empire* (Cambridge: Harvard University Press, 2012).

18. Historian Philip Scranton describes power carpet looms as "double-edged innovations" that led to job loss and price-damaging overproduction. Scranton, "Build a Firm, Start Another: The Bromleys and Family Firm Entrepreneurship in the Philadelphia Region," *Business History* 35, no. 4 (Winter 1993): 117. For the displacement of less technologically adaptive artisan-entrepreneurs by the shift to mechanization, see David Jaffee, *A New Nation of Goods: The Material Culture of Early America* (Philadelphia: University of Pennsylvania Press, 2010).

19. Walter Licht, *Industrializing America: The Nineteenth Century* (Baltimore: Johns Hopkins University Press, 1995), 34. For the role of the South in the rise of industrial capitalism, see Walter Johnson, *River of Dark Dreams: Slavery and Empire in the Cotton Kingdom* (Cambridge: Harvard University Press, 2013). For the connections between northern factories and southern plantations, see Seth Rockman, *Plantation Goods and the National Economy of Slavery in Antebellum America* (Chicago: University of Chicago Press, forthcoming).

20. Licht, *Industrializing America,* 35.

21. I am defining "big business" as corporations that depend on complex supply and distribution networks coordinated by managers and large capital markets to raise funds ahead of revenue. Jack High, "Economic Theory and the Rise of Big Business in America, 1870–1910," *Business History Review* 85, no. 1 (Spring 2011): 86–87.

22. Carl Seaburg and Stanley Paterson, *Merchant Prince of Boston: Colonel T. H. Perkins, 1764–1854* (Cambridge: Harvard University Press, 1971), 344.

23. Richard White, *Railroaded: The Transcontinentals and the Making of Modern America* (New York: W. W. Norton, 2011).

24. Licht, *Industrializing America,* 98.

25. Wilson, *The Business of Civil War,* 225.

26. Samuel Goodrich, *Supplement to the Pictorial Geography of the World, for 1841 and 1842* (Boston: C. D. Strong, 1842), 31.

INDEX